ENGLISH AND ETHNICITY

Signs of Race

General Editors: Phillip D. Beidler and Gary Taylor

Writing Race across the Atlantic World: Medieval to Modern, edited by Phillip D. Beidler and Gary Taylor (January 2005)

Buying Whiteness: Race, Culture, and Identity from Columbus to Hip-Hop, by Gary Taylor (January 2005)

English and Ethnicity, edited by Janina Brutt-Griffler and Catherine Evans Davies (December 2006)

ENGLISH AND ETHNICITY

Edited by
Janina Brutt-Griffler
and
Catherine Evans Davies

ENGLISH AND ETHNICITY

First published in 2006 by
PALGRAVE MACMILLAN™
175 Fifth Avenue, New York, N.Y. 10010 and
Houndmills, Basingstoke, Hampshire, England RG21 6XS
Companies and representatives throughout the world.

PALGRAVE MACMILLAN is the global academic imprint of the Palgrave Macmillan division of St. Martin's Press, LLC and of Palgrave Macmillan Ltd. Macmillan® is a registered trademark in the United States, United Kingdom and other countries. Palgrave is a registered trademark in the European Union and other countries.

ISBN-13: 978–0–312–29599–8 (hardcover)
ISBN-10: 0–312–29599–5 (hardcover)
ISBN-13: 978–0–312–29600–1 (paperback)
ISBN-10: 0–312–29600–2 (paperback)

Library of Congress Cataloging-in-Publication Data is available from the Library of Congress.

A catalogue record for this book is available from the British Library.

Design by Newgen Imaging Systems (P) Ltd., Chennai, India.

First edition: December 2006

10 9 8 7 6 5 4 3 2 1

Printed in the United States of America.

CONTENTS

CONTRIBUTORS

Dr. John Baugh, Director of African and Afro-American Studies; Departments of Psychology, Anthropology, Education, and English, Washington University in St. Louis, USA. Known for educational and social applications of linguistic science, Dr. Baugh taught for 14 years at Stanford University before becoming the inaugural holder of an endowed professorship. While at Stanford, Dr. Baugh was Director of the Stanford Teacher Education Program from 1994–1996, and he received Stanford University's St Clair Drake Teaching Award in 1999–2000. He was the president of the American Dialect Society from 1992–1994, and has served on advisory committees for collegiate dictionaries. His most recent books include *Beyond Ebonics: Linguistic Pride and Racial Prejudice (2000), and Out of the Mouths of Slaves: African American Language and Educational Malpractice* (1999).

Dr. Cynthia Goldin Bernstein, Department of English, University of Memphis, USA. Dr. Bernstein taught for ten years at Auburn University, where she co-organized the second decennial symposium on Language Variety in the South, and coedited the resulting volume *Language Variety in the South Revisited* (1997). She currently teaches in a new Ph.D. program in language study. Dr. Bernstein's research interests include the study of dialect variation, methods of gathering and interpreting linguistic data, and the use of linguistic tools to analyze literary and popular genres. She is a past president of the Southeastern Conference on Linguistics. Her publications include *The Text and Beyond: Essays in Literary Linguistics* (edited, 1994).

Dr. Janina Brutt-Griffler, Department of Learning and Instruction, State University of New York at Buffalo, USA. Prior to assuming her current position, Dr. Brutt-Griffler taught on the graduate faculty at the University of York, England. At the time of the symposium from which the current volume is drawn, she was an Associate Professor in the Department of English at the University of Alabama. She is the author of *World English: A Study of Its Development* (2002), winner of the Modern Language Association's 2004 Kenneth W. Mildenberger

Prize. Her other publications include *Bilingualism and Language Pedagogy* (coedited 2004). She serves on the Executive Committee for General Linguistics of the Modern Language Association and the Distinguished Service and Scholarship Committee of the American Association for Applied Linguistics.

Dr. A. Suresh Canagarajah, Department of English, Baruch College, City University of New York, USA. Dr. Canagarajah's research interests span bilingualism, discourse analysis, academic writing, postcolonial literature, and critical language pedagogy. His book, *Resisting Linguistic Imperialism in English Teaching* (1999), was awarded the Mina P. Shaughnessy Award (2000) by the Modern Language Association for the best "research publication in the field of teaching English language, literature, rhetoric, and composition." He has most recently edited a collection of articles by international scholars on responses to globalization in *Reclaiming the Local in Language Policy and Practice* (2005). He serves currently as editor of the *TESOL Quarterly*, the flagship journal of the international association for applied linguists.

Dr. Nikolas Coupland, Professor and Director of the Cardiff Centre for Language and Communication Research, University of Cardiff, Wales, United Kingdom. Dr. Coupland is the founding editor (with Allan Bell) of the *Journal of Sociolinguistics*. His primary areas of research include sociolinguistics and sociolinguistic theory, style variation in spoken language, the sociolinguistics of Wales, critical discourse analysis, and lifespan communication. He currently directs the Centre's project on Welsh language and Welsh identity under globalisation. His books include: *Researching Language Attitudes: Social Meanings of Dialect, Ethnicity and Performance* (coedited, 2003), *Sociolinguistics and Social Theory* (coedited, 2001), *Discourse and Lifespan Identity* (1993, with Jon Nussbaum), *Language: Contexts and Consequences* (1991, with Howard Giles), *English in Wales* (1990), and *Dialect in Use: Sociolinguistic Variation in Cardiff English* (1988).

Dr. Catherine Evans Davies, Department of English, University of Alabama USA. At the time of the symposium from which the current volume is drawn, Dr. Davies was director of the graduate linguistics programs in the English Department at the University of Alabama. Her work on cross-cultural interaction has appeared in *World Englishes, Text, Multilingua, the Journal of Pragmatics*, and elsewhere. She served as president of the Southeastern Conference on Linguistics

in 2003–2004. With Michael D. Picone she was co-organizer of the
NSF-funded decennial symposium on "Language variety in the south
historical and contemporary perspectives," in 2004, and is currently
coediting the resulting volume under an NEH grant.

Dr. Marcia Farr, Departments of Education and English, Ohio State
University; Professor Emerita of English and Linguistics, University
of Illinois at Chicago, USA. As a sociolinguist, Dr. Farr focuses on
cultural variation in the use of oral and written language and the ways
this affects teaching and learning. Her current research is a long-term
ethnographic study of culture, language and identity among a
transnational social network of Mexican families in Chicago and their
village-of-origin in Michoacan, Mexico. She has edited two volumes
on this topic *Ethnolinguistic Chicago: Language and Literacy in the
City's Neighborhoods* (2004), and *Latino Language and Literacy in
Ethnolinguistic Chicago* (2005). She has a forthcoming edited volume
entitled *Rancheros in Chicagoacán: Ways of Speaking and Identity in a
Mexican Transnational Community*.

Dr. Yunte Huang, Department of English, University of California at
Santa Barbara, USA. Dr. Huang received his B.A. from Peking
University; his M.A. from the University of Alabama; and his Ph.D.
from SUNY-Buffalo. After teaching at Harvard University, Dr. Huang
became director of the Consortium for Literature, Theory, and Culture
at UC Santa Barbara. His publications include: *Transpacific
Displacement: Ethnography, Translation, and Intertextual Travel in
Twentieth-Century American Literature* (2002); *Shi: A Radical
Reading of Chinese Poetry* (1997), *Selected Language Poems* (1993), and
a translation into Chinese of Ezra Pound's *The Pisan Cantos*. Both a
scholar and a poet, Dr. Huang is an affiliated faculty member of UCSB's
interdisciplinary American Cultures and Global Contexts Center.

Dr. Alamin A. Mazrui, Department of African-American and African
Studies, Ohio State University, USA. Dr. Mazrui has taught at
universities in Kenya and Nigeria, and has served as a consultant to
nongovernmental organizations in Africa on language and urbaniza-
tion, and language and the rule of law. His latest publications include,
English in Africa: After the Cold War (2004), and *Power of Babel:
Language and Governance in the African Experience* (1998), coauthored
with Ali A. Mazrui. Dr. Mazrui is a published playwright and poet in
Kiswahili.

Professor Simon J. Ortiz, Department of English, University of
Toronto, Canada. Professor Ortiz is a member of Acoma Pueblo in

New Mexico. "I've been a poet for thirty years," he writes, "mainly trying to demystify language and enhance its meaning for me and readers and listeners." The author of 19 books of poetry and prose, he says that most of his work "focuses on issues, concerns and responsibilities we, as Native Americans, must have for our land, culture and community." His most recent book of poems is *From Sand Creek: Rising in This Heart Which is Our America* (2000). He is coeditor of *Beyond the Reach of Time and Change: Native American Reflections on the Frank A. Rinehart Photograph Collection* (2005).

Dr. Donna Patrick, Department of Sociology and Anthropology and the School of Canadian Studies, Carleton University, Canada. Dr. Patrick's research focuses on the political, social, and cultural aspects of language use, community practices, and nation-building among Aboriginal peoples in Canada, with a primary focus on Inuit (primarily in Northern Quebec and in urban centres). Interests include the relation between language and power, especially as this involves minority language rights and policy; and the interrelations between language and political economy, ideology, and globalization. Recent publications include *Language Rights and Language Survival* (coedited, 2004), *Language, Politics, and Social Interaction in an Inuit Community* (2003), and *Linguistic Minorities and Modernity. A Sociolinguistic Ethnography* (collaboration with Monica Heller, Phyllis Daley, and Mark Campbell, 1999).

Dr. John R. Rickford, Department of Linguistics, Director of African and Afro-American Studies, Stanford University, USA. Dr. Rickford's research interests include sociolinguistics, especially the relation between language and ethnicity, social class and style, language variation and change, pidgin and creole languages, African American Vernacular English, and the applications of linguistics to educational problems. His books include *A Festival of Guyanese Words* (edited, 1978), *Dimensions of a Creole Continuum* (1987), *Analyzing Variation in Language* (coedited, 1987), *Sociolinguistics and Pidgin-Creole Studies* (edited, 1988), *African American English* (coedited, 1998), *Creole Genesis, Attitudes and Discourse* (coedited, 2000), *Style and Sociolinguistic Variation* (coedited, 2002). *Spoken Soul: The Story of Black English* (coauthored with his journalist son, 2000) was the winner of an American Book Award.

Dr. Almeida Jacqueline Toribio, Department of Spanish, Italian and Portuguese, Pennsylvania State University, USA. Dr. Toribio's research explores language in its social and cultural context;

combining linguistic-theoretical approaches with insights from the fields of psychology, education, and sociology. At the 4th International Symposium on Bilingualism, she convened a session on formal perspectives on bilingual codeswitching. In addition to articles that adopt a more formal perspective, her publications include a review of *Literacy Development in a Multilingual Context: Cross-Cultural Perspectives* (*Language*, 2001), "Language variation and the linguistic enactment of identity among Dominicans" (*Linguistics*, 2000), "Review article: research directions in second language acquisition" (*World Englishes*, 2000).

INTRODUCTION

Catherine Evans Davies

An important organizing principle of the symposium from which this volume has emerged was diversity of voices, expressing different disciplinary orientations, contexts under study, theoretical frameworks, methodologies, and perspectives on the topic. The disciplinary orientations represented include not only applied (socio)linguistics, but also literary studies, cultural anthropology, communication, African American studies, and education. The contexts under study are intra-national, transnational, international, and global. The relation of English and ethnicity is examined not only among American subcultural groups such as African Americans and Jewish Americans, but also among such diverse groups as Native Americans (both within the United States and Canada), the Welsh in Britain, and Africans. It is further explored among Mexicans who travel between Mexico and the United States, and with Dominicans and Sri Lankan Tamils in diaspora. It is extended transculturally to Chinese. Theoretical frameworks range from a micro-sociolinguistic focus on contextualized discourse analysis, to a macro-sociolinguistic treatment with social-psychological dimensions, to a historical crosscultural literary focus with an autobiographical component. All are concerned, whether explicitly or implicitly, with the relation between language and culture. In terms of methodology, a majority of the essays have an ethnographic dimension. Most present data in some form, whether from recordings gathered during field work, or from literature, or from popular culture.

The perspectives on the topic are myriad, and we encouraged a range of perspectives through the coordinated structure of our title, "English and Ethnicity." The volume includes professional academics writing about both their own cross-cultural linguistic experience and that of groups to which they bring an outsider's perspective. We also hear advocacy from a poet for his endangered language and cultural perspective. The volume includes a political perspective on language and the schools in the United States, as well as documentation of various linguistic effects of globalization. Represented are current perspectives on an ancient colonial context as well as modern post-colonial contexts. Subcultures within the United States receive attention in terms of links between language and culture, linguistic representation in literature, and the negotiation of identity. The contributors' individual styles are represented on the written page, but a fascinating dimension of the live symposium is missing: their diverse accents in English.

What unites the essays in this volume is their orientation to the stated theme of the symposium, a theoretical position on English and ethnicity that challenges certain traditional assumptions and frameworks. The participants in the symposium responded to the following statement:

> Our focus in this symposium will be the use of English as a resource for the representation of ethnicity as an aspect of sociocultural identity. Our theoretical position is that ethnicity is potentially an aspect of the identity of every person, and that English can be used to signal a wide range of ethnicities in a wide range of contexts. Such a position problematizes certain key notions: the notion of identity must be conceptualized as complex, multifaceted, and socially constructed through a process of situated interpretation; the notion of ethnicity must be conceptualized as both subsuming and transcending earlier notions of "race" as well as including a wide range of perceptions of relevant cultural background; English itself must be conceptualized not as a monolithic linguistic entity with one "standard" form, but as a highly complex linguistic construct with spoken and written forms, and a wide range of dialectal variation that can be conveyed through shifts at all levels of linguistic organization (prosodic, phonological, lexical, morpho/syntactic, pragmatic, discoursal). The symposium will include papers which address regional, national, and international contexts in the exploration of the relationship between English and ethnicity.

Contrasting this volume with a recent landmark publication in the general field of "language and ethnic identity" (Fishman 1999), several significant differences emerge. The first difference is our inclusion, among disciplinary perspectives, of that of literary and cultural studies as represented in the essays by Ortiz, Huang, and Bernstein. The second is our attempt to move beyond the treatment of languages (or even dialects) as monolithic. Whereas the Fishman volume acknowledges the complexity of the notions of both identity and ethnicity, it still appears to treat languages, in discussions of "region and language perspectives," as uniform entities. Such a treatment precludes the kind of analysis called for here. Finally, in a difference that flows from our refusal to treat "English" as monolithic, the current volume includes a discourse perspective, with discourse data as primary in a number of the contributions. In the Fishman volume, in contrast, the only papers in which language data appear (Lanehart 1999; McCarthy and Zepeda 1999) are those in which the data are used to illustrate language attitudes in relation to identity, but not to provide evidence for how speakers actually use English in interaction as a resource for the representation of ethnicity as an aspect of sociocultural identity.

The general orientation to the symposium arose partly from the recognition that the discipline of sociolinguistics embraces a range of approaches and that current thinking encourages diversity (Coupland et al. 2001; Eckert and Rickford 2002). Another consideration was the position of the organizers as linguists in an English Department. Within this context it is important to demonstrate important links between linguistics and literary and cultural studies, and to establish the relevance of linguistic analysis, and of sociolinguistics in particular, to the study of literature. The location of the symposium at the University of Alabama also played a role in that local issues of language and ethnicity needed to be addressed. The first and most obvious local concern is African American Vernacular English. Relevant contributions to the volume are Mazrui in relation to Afrocentricity and Pan-Africanism, Baugh in relation to the sociopolitical context of contemporary public education, and Rickford in relation to historical cultural continuity between contemporary African Americans, the Caribbean, and Africa. A second local concern is the growing population of Latinos in rural Alabama. Enlightening contributions for Alabamians seeking to understand more about the situation are Farr's description of the transnational experience of some Mexican immigrant populations and their struggle with ethnic categorization and language, and Toribio's exploration of the phenomenon of code switching in a Spanish-speaking immigrant population in relation to language attitudes and prejudice. In addition, the organizers felt that the local audience needed to hear diverse voices from other contexts, in order to locate their particular circumstances within a broader perspective. For example, Patrick's work examines a multilingual context that might seem exotic to Alabamians but in fact represents a common experience in the rest of the world. As Alabama responds to globalization by hosting communities associated with industries from Germany and Japan, Huang highlights the international importance of the English language in a particular context, that of China. Both Ortiz and Patrick show the struggle of indigenous people in North America to maintain and perpetuate bilingualism in order to retain their ancestral languages while also learning English as a tool in the modern world. Both of these papers offer an enlightening perspective on current ideological struggles with notions of "bilingual education."

Finally, a significant goal of the symposium was to attract audiences from related disciplines in order to demonstrate that linguists' concerns and methodologies have potential relevance for their own work. Our contributors themselves come from departments of English, Communication Research, Linguistics, Cultural Anthropology, Sociology, Psychology, Education, Learning and Instruction, African and Afro-American Studies, American Studies, Canadian Studies, and Romance Languages (Spanish, Italian, and Portuguese). Our contributors include not only scholars concerned with language, but also poets, fiction writers, and a playwright. Sponsors for the symposium, as indicated more fully in the acknowledgements, included the departments of Anthropology, History, Psychology, Religious Studies, Modern Languages and Classics, and American Studies. Additional sponsors were the Creative Writing Program, Capstone International Programs, and the English Language Institute, as well as the College of Education. Through these sponsorships we established interdisciplinary links on our campus, and through the sponsorship of Stillman College we created a link with a historically black institution in Tuscaloosa. Thus the intended audience for this volume is diverse.

CHALLENGES TO CERTAIN TRADITIONAL ASSUMPTIONS AND FRAMEWORKS

The diversity of approaches is significant in that the stated theoretical position represents a challenge to certain traditional conceptualizations of the notions of identity, ethnicity, and language. Taking the perspective that language is a resource for the representation of ethnicity as an aspect of sociocultural identity, and that ethnicity is potentially an aspect of the identity of everyone, problematizes certain key notions. The notion of identity, rather than essentialist or fixed, must be conceptualized as complex, multifaceted, and socially constructed through a process of situated interpretation. Whereas the complexity of the notion may be recognized in other work, it is not typically treated as an interactional accomplishment. Defining language as a "resource" shifts perspective to the subjectivity and agency of the individual speaker who, though working within the conceptual cultural universe inscribed in language, is not simply blindly reproducing it but rather using language selectively and strategically for purposes of individual self-presentation and then negotiating aspects of identity in interaction with other speakers. Ethnicity, rather than being somehow synonymous with "race," is potentially an aspect of the identity of every person.

The notion of ethnicity must be conceptualized as both subsuming and transcending earlier notions of "race" as well as including a wide range of perceptions of relevant cultural background. Even though the linguist organizers agreed to place the symposium under the larger rubric of a series called "Signs of Race," we resisted using the term "race" in the title of our volume and opted instead for "ethnicity." Whereas linguists recognize that it is important to interrogate the naturalized notion of "race" and to deconstruct it, as was done in the previous symposium (Beidler and Taylor 2005), they are highly sensitive both to the power of language to reify concepts, and also to the widely held folk-linguistic assumption that race and language are somehow related.

Finally, English itself must be conceptualized not as a monolithic linguistic entity with one "standard" form, but as a highly complex linguistic construct with spoken and written forms, and a wide range of dialectal variation that can be conveyed through shifts at all levels of linguistic organization (prosodic, phonological, lexical, morpho/ syntactic, pragmatic, discoursal), that is to say, through intonation patterns, accent, choice of words, sentence patterns, and distinctive ways of using language to, for example, convey politeness (C. Davies 2002, 2003). English can be used to signal a wide range of ethnicities in a wide range of contexts, and no one feature or variable has an inherent semiotic value. Such a perspective, of course, recognizes not only the importance of culture as context (C. Davies 1998), but also the reflexive relationship between context and language.

A "SOCIOLINGUISTIC TURN"

Since the volume is designed to attract scholars of literary and cultural studies among its readers, it seems appropriate to take the opportunity to comment on the recent use of the term "linguistic" within that disciplinary area and to clarify the position taken by the organizers as linguists. What has been called "the linguistic turn" within literary and cultural studies appears to have taken several twists within the turn. The first twist, generally speaking, pursued Saussurean insights into a world in which the "subject" is created by the linguistic context and thus relinquishes agency to discourse.

A more recent twist, as described in Young (2001) in terms of a focus within African American studies on "narratives of specificity," appears to isolate language within an updated version of "New Criticism," and seems to represent a turning away from contextualization to an intentionally apolitical and radically subjective stance. Whereas we recognize that linguistics as practiced within the generativist paradigm may indeed abstract from context (Chomsky 1965), sociolinguistic traditions within the discipline have been dedicated to conceptualizing and demonstrating the relations between language and context (narrowly or broadly defined). The dominant variationist tradition, it is true, has tended to use positivist frameworks to establish correlations between linguistic features and social categories (Chambers 2003). Fought (2002), however, in an entry on "ethnicity" in a current handbook of language variation and change, though limiting her discussion to variationist studies, does focus on the use of linguistic variables by speakers as "acts of identity." Such variables, however, are traditionally restricted to phonology, morphology, and syntax.

The "sociolinguistic turn" represented in this volume, in contrast, is to a broader and more nuanced and radically contextualized discourse-based view of language as embedded in culture. Language conveys not only referential meaning but also social meaning. In this view of language, positivist frameworks and methodologies no longer seem to apply. If it is difficult to pin down identity or ethnicity or "English" or aspects of context to be identified as variables, then a methodology that attempts to correlate them in the tradition of quantitative variationist sociolinguistics seems inappropriate and inadequate. If the language itself, within this set of redefinitions, reflexively creates context through indexicality (Duranti and Goodwin 1992), then discourse analysis as a methodology becomes imperative. If the choice of language is an individual (but socioculturally constrained) deployment of resources in a context, then subjective considerations not only of attitude (as traditionally measured through surveys and interviews) but of presentation of self in moment-to-moment social interaction become essential to the analysis. The whole question of subjectivity requires qualitative methodologies, based on the ethnographic approaches of anthropology toward understanding language and culture, in order to examine individual interpretations in interaction (Johnstone 1996, 2000). These qualitative methodologies can of course be supplemented with appropriate quantitative methodologies.

LINKING INDIVIDUAL LANGUAGE
USE TO SOCIAL ORGANIZATION

The question of the linking of individual behavior to social organization (conceptualized as "larger social structures" in the form of institutions) has been a perennial problem within the social sciences, and thus within sociolinguistics as it exists with one foot in the humanities and the other in the social sciences (Coupland et al. 2001). As Sapir (1931) expressed the relationship, "[society] is only apparently a static sum of social institutions; actually it is being reanimated or creatively reaffirmed from day to day by particular acts of a communicative nature which obtain among individuals participating in it." The key question, of course, is the relationship between the social constraints as shapers of those individual acts of a communicative nature and the freedom of the individual speaker to act outside of those social constraints for her or his own purposes. A clue to Sapir's attitude may lie in his choice of the word "creatively" in his use of the paired terms, "reanimated or creatively reaffirmed."

TOP-DOWN APPROACHES, EXPANDED

One direction of theorizing and research has been a top–down approach of starting with social categories or structures, locating individuals within these categories or structures (according to some predetermined "objective" criteria), and then assuming a sort of determinism that might predict their behavior, linguistic or otherwise. The linguistic reflex of this has been to identify "dialects" or "varieties"—in the case of dialects, by certain recurring linguistic features of phonology, morphology, lexicon, and/or syntax that may deviate from a "standard"—and then to identify individuals as speakers of the dialect if they employ this set of features. Such an approach encounters problems when individuals use only a subset of features or select a feature that has particular symbolic value for the purpose of "crossing" (Bucholtz 2002; Rampton 1995) or otherwise performing their identity or ethnicity (C. Davies in press; Dubois and Horvath 2002; Johnstone 1999). The approach is linked to certain ideologies within the discipline that are rooted in the Romanticism of early dialectology that was searching for the "authentic" and "pure" speaker of a dialect, preferably archaic (Bucholtz 2002).

Such an approach may have been valuable in an idealized world of stable communities, but in a postmodern, globalized world in which the subjectivity and agency of speakers are a significant aspect of their sociolinguistic performance, it can no longer apply. In a world in which mother-tongue speakers of English are outnumbered by speakers of English as an additional language, the category of "native speaker" needs to be interrogated in the same way (A. Davies 1991, 2003). Indeed, Goffman's (1959, 1974, 1981) work on the presentation of self in everyday life serves as a touchstone, broadened and refined to deal with language variation (Coupland 2002; C. Davies 2002). Another important extension is from the restriction of the "traditional" linguistic levels (of phonology, morphology, and syntax) to a more inclusive framework (C. Davies 1997) that incorporates pragmatics and discourse conventions as an aspect of language variety.

BOTTOM-UP APPROACHES, EXPANDED

The complementary bottom–up approach to the linking of individual behavior and social structure (represented most clearly by "conversation analysis," for example, Sacks et al. 1974) has been to do microanalysis of interaction and allow social categories to emerge only through the data, rigorously excluding subjective interpretations by the participants. The assumption is that the relevant categories will emerge as what the interactants are orienting to within the discourse. Less restricted versions of this approach (rooted in cognitive sociology as exemplified in the work of Cicourel (1978)) and in anthropology in the work of Erickson and Shultz (1982)—and returning to the perspective of Sapir quoted above) take the microinteractions as the core data of social reality from which the larger social groupings and institutions are constructed, and require subjective self-categorization as an important dimension of methodology. Thus it is not only how speakers are categorized by "objective" criteria that has significance for language use, but also how they categorize themselves in accordance with how they choose to act (or attempt to act). This expanded framework is represented by work based in Gumperz (1982, 1992), such as Heller and Martin-Jones (2001). It moves toward a different notion of how speakers are grouped, toward the idea of interpretive communities with an intersubjectivity that entails shared interpretive conventions.

FROM "SPEECH COMMUNITY" TO "COMMUNITY OF PRACTICE" WITH SHARED INTERPRETIVE CONVENTIONS

Such a movement involves a radical shift from the traditional idea of "speech community," problematic though that concept has been (cf. Patrick 2002). It no longer makes sense to think in terms of a speech community as a linguistic object, whether homogeneous (Chomsky 1965) or heterogeneous (Hymes 1974), within which an individual is located according to knowledge of a particular code. Neither does it make sense to think of a speech community as a purely social object, within which an individual is located according to some "objective" criteria. The shift needs to be to a grouping that has an essential subjective component, that of the individual's ability to understand social interaction and to interact effectively within a certain group. The notion of "discourse community" has been proposed within applied linguistics (Swales 1990), but the designation may still allow an approach using so-called objective criteria that falls into the same trap as "speech community." A clearer shift is indicated by the term "interpretive community," if defined by Gumperz's (1982) notion of shared interpretive conventions. The term "interpretive community" was actually coined within literary theory by Fish (1980) to explain different readings of a text, but the literary formulation is much more deterministic than would be compatible with a vision of individual speaker agency that is intended here.

Shared modes of interpretation in Gumperz's sense may coincide with traditional social categories, or they may cut across traditional social categories of class, race, gender, or ethnicity. Further, shared modes of interpretation may not be coterminous with knowledge of a language as traditionally defined. Thus, a speaker may share "English" as a code, but not a subset of conventions for signaling and interpreting a particular ethnicity. Thus there can be no essential semiotic value for a particular feature or variable, but rather only a relational meaning. Shifting to the idea of "interpretation" rather than, for example, "linguistic proficiency" also opens up the possibility of receptive competence rather than concentrating solely on the ability to produce language. Thus exposure to a range of varieties of English through the mass media may develop receptive competence in a wide range of interpretive communities, but not necessarily the ability to interact effectively within that interpretive community.

The notion of "community of practice" (Lave and Wenger 1991; Wenger 1998) is gaining currency within sociolinguistics (Holmes and Meyerhoff 1999) as an alternative to the notion of speech community, among others. The key idea is that there is a common enterprise, a qualitative distinction from other models that focus on quantity of contact. Eckert (2000, 35) points out that "the value of the construct *community of practice* is in the focus it affords on the mutually constitutive nature of individual, group, activity, and meaning." Whereas there are degrees of membership in the community of practice, it is potentially implicit that the community involves shared "interpretive practices" (Gumperz and Levinson 1996). Holmes and Meyerhoff (1999, 181) and Bucholtz (1999, 210) both see the community of practice framework as having more potential for linking micro-level and macro-level analyses. Unlike the speech community construct that is constructed from analysts' categories, the community of practice construct, through its dependence on ethnographic methods, includes practitioners' categories in a meaningful way. The essay by Canagarajah in this volume makes use of the notion of community of practice in order to explain language shift in relation to diasporic identities.

INDIVIDUAL SUBJECTIVITY AND AGENCY

To conceptualize language as a "resource" emphasizes the agency and subjectivity of the individual speaker. Given the sociocultural constraints of a particular context, the speaker is not simply acting as a social automaton, but is rather to some extent "creatively reaffirming" social organization through purposive deployment of language. Variationist sociolinguistic studies are moving in this direction as represented by the focus of Fought (2002). She chooses to discuss "ethnicity" within a current handbook by selecting variationist studies that frame the use of variables in terms of "acts of identity" (LePage and Tabouret-Keller 1985) and that begin to explore the notion of "crossing" in which a speaker uses variables from the dialect of the Other. Schilling-Estes (2004) finds that "ethnic varieties—and ethnic identities themselves—are not neatly bounded, monolithic entities but rather that different people—and peoples—freely adopt and adapt linguistic and cultural resources from one another, both at the local level, in unfolding interaction, and on a more global level, in shaping and reshaping group varieties over time and across space." Thus we see the complex interrelationship between the individual speaker and social organization. In Coupland's essay the elaboration of the notion of social context allows a closer and more nuanced link between individual behavior, conceptualized as "the situated, dynamic and strategic projection of social identities," and larger social structures.

THE SOCIAL CONSTRUCTION OF
ETHNIC IDENTITY IN INTERACTION

Finally, this shift to a radically contextualized discourse analysis also entails the idea that ethnicity as an aspect of identity is socially constructed in interaction (Ochs 1992), signaled linguistically through what Gumperz (1982, 1992) has termed "contextualization cues," constellations of features at different levels of linguistic organization that convey, within the interpretive community, social meaning. The essay by Coupland uses discourse data to show how phonological features serve as contextualization cues. Further, he shows how the signaling through those features shifts within the discourse in relation to the persona that the speaker is intending to convey with attendant layers of social meaning in the performance context. The links between the linguistic features and the social meanings are necessarily indirect. These meanings can be multiple in relation to different discourse frames, and they must be taken up (by an audience or an interlocutor) in order for the identity to be ratified, or jointly constructed (C. Davies 2005). The hermeneutic quality of the analysis required, drawing on inferences signaled by different contextualization cues, and ideally involving multiple perspectives as part of the analysis, has a literary feel.

Diversity but not Necessarily Hybridity

The contributions to this volume reflect diversity in conformity with the theoretical perspective outlined above. The volume that results is not intended to represent hybridity, but rather to offer different voices, each with a different orientation to the general theme. Given the complexity of the topic, a range of approaches need to be explored, and massive amounts of data need to be collected and analyzed, before a hybrid framework can be conceptualized, one that allows us to effectively link individual linguistic behavior and social organization as mediated through language. It is clear, however, that English serves in many complex ways as a resource for the representation of ethnicity as an aspect of sociocultural identity. It is also clear that is it simultaneously both a unifying and a diversifying force.

Acknowledgments

We gratefully acknowledge support from the College of Arts & Sciences, the Provost, the Dean of Arts & Sciences, the Arts & Sciences Diversity Committee, the College of Education, the Department of American Studies and the African-American Studies Program, Capstone International Programs, the Department of Religious Studies and the Aaron Aronov Endowment for Judaic Studies, the

Creative Writing Program, the History Department, the Psychology Department, the Modern Languages and Classics Department, the Anthropology Department, the English Language Institute, Stillman College, and the Alabama Humanities Foundation, the state affiliate of the National Endowment for the Humanities.

The editors of the volume also gratefully acknowledge the invaluable support of our linguist colleague, Dr. Lucy Pickering, in the organizing of the symposium, particularly her efficient handling of the finances. We also wish to acknowledge the essential help of our graduate students. Rachel Shuttlesworth designed our posters and brochures, and organized the appropriate technology for the conference presenters. Kelli Loggins Keener coordinated transportation, both between Tuscaloosa and the Birmingham Airport and also for local shuttles for participants. Faizah Sari saw to it that all of our sessions were videotaped, and organized a display of participants' publications in the library. Ramona Hyman, Janice Filer, Tasheka Gipson, Kirsten Fidel, Melissa Chase, Jong Hee Shadix, Sylvia Koestner, Jeong Hee Ahn, Christy Beem, Josie Prado, Christina Womack, and Frannie James all helped with registration and in myriad other ways.

REFERENCES

Beidler, Phillip D. and Gary Taylor (eds.). 2005. *Writing Race across the Atlantic World: Medieval to Modern*. New York: Palgrave.

Bucholtz, Mary. 1999. "Why be normal?" Language and identity practices in a community of nerd girls. *Language in Society* 28, 2: 203–223.

Chambers, J.K. 2003. *Sociolinguistic Theory* (2nd edition). Oxford: Blackwell Publishing.

Chomsky, Noam. 1965. *Aspects of the Theory of Syntax*. Cambridge, MA: MIT Press.

Cicourel, Aaron. 1978. Language and society: Cognitive, cultural and linguistic aspects of language use. *Sozialwissenschaftliche Annalen* 2: 25–58.

Coupland, Nikolas. 2002. Language, situation, and the relational self: Theorizing dialect-style in sociolinguistics. In *Style and Sociolinguistic Variation*, P. Eckert and J.R. Rickford (eds.). Cambridge: Cambridge University Press. 185–210.

Coupland, Nikolas, Srikant Sarangi, and Christopher N. Candlin (eds.). 2001. *Sociolinguistics and Social Theory*. New York: Longman.

Davies, Alan. 1991. *The Native Speaker in Applied Linguistics*. Edinburgh: The University of Edinburgh Press.

Davies, Alan. 2003. *The Native Speaker: Myth and Reality*. Clevedon: Multilingual Matters.

Davies, Catherine Evans. 1997. Social meaning in Southern speech from an interactional sociolinguistic perspective: An integrative discourse analysis of terms of address. In *Language Variety in the South Revisited*, C. Bernstein, T. Nunnally, and R. Sabino (eds.). University of Alabama Press. 225–241.

Davies, Catherine Evans. 1998. Maintaining American face in the Korean oral exam: Reflections on the power of crosscultural context. In *Talking and Testing: Discourse Approaches to the Assessment of Oral Proficiency*, R. Young and A.W. He (eds.). John Benjamins. 271–296.

Davies, Catherine Evans. 2002. Martha Stewart's linguistic presentation of self. *Texas Linguistic Forum* 44, 1: 73–89.

Davies, Catherine Evans. 2003. Martha Stewart and American "Good Taste." In *New Media Language*, Jean Aitchison and Diana Lewis (eds.). Routledge. 146–155.

Davies, Catherine Evans. 2005. Learning the discourse of friendship. In *Language in Use: Cognitive and Discourse Perspectives on Language and Language Learning*, Andrea Tyler et al. (eds.). Washington, DC: Georgetown University Press (Georgetown University Round Table on Languages and Linguistics). 85–99.

Davies, Catherine Evans. In press. Language and identity in discourse: Sociolinguistic repertoire as expressive resource in the presentation of self. In Talk and Identity in Narratives and Discourse, A. De Fina, M. Bamberg, and D. Schiffrin (eds.). Amsterdam: Benjamins.

Dubois, Sylvie and Barbara Horvath. 2002. Sounding Cajun: The rhetorical use of dialect in speech and writing. *American Speech* 77, 3(Fall): 264–287.

Duranti, Alessandro and Charles Goodwin (eds.). 1992. *Rethinking Context: Language as an Interactive Phenomenon*. Cambridge: Cambridge University Press.

Eckert, Penelope. 2000. *Linguistic Variation as Social Practice*. Oxford: Blackwell.

Eckert, Penelope and John R. Rickford (eds.). 2002. *Style and Sociolinguistic Variation*. Cambridge: Cambridge University Press.

Erickson, Frederick and Jeffrey Shultz. 1982. *The Counselor as Gatekeeper: Social Interaction in Interviews*. New York: Academic Press.

Fish, Stanley E. 1980. *Is There a Text in This Class? The Authority of Interpretive Communities*. Cambridge: Harvard University Press.

Fishman, Joshua (ed.). 1999. *Handbook of Language and Ethnic Identity*. New York: Oxford University Press.

Fought, Carmen. 2002. Ethnicity. In *The Handbook of Language Variation and Change*, J.K. Chambers, Peter Trudgill, and Natalie Schilling-Estes (eds.). Oxford: Blackwell. 444–472.

Goffman, Erving. 1959. *The Presentation of Self in Everyday Life*. New York: Doubleday.

Goffman, Erving. 1974. *Frame Analysis*. New York: Harper and Row.

Goffman, Erving. 1981. *Forms of Talk*. Philadelphia: University of Pennsylvania Press.

Gumperz, John J. 1982. *Discourse Strategies.* Cambridge: Cambridge University Press.

Gumperz, John J. 1992. Contextualization cues, In *Rethinking Context: Language as an Interactive Phenomenon,* Alessandro Duranti and Charles Goodwin (eds.). Cambridge: Cambridge University Press. 229–252.

Gumperz, John J. and Stephen Levinson. 1996. Introduction: Linguistic relativity re-examined. In *Rethinking Linguistic Relativity,* J.J. Gumperz and S.C. Levinson (eds.). Cambridge: Cambridge University Press. 1–18.

Heller, Monica and Marilyn Martin-Jones (eds.). 2001. *Voices of Authority: Education and Linguistic Difference.* Westport, CT: Ablex.

Holmes, Janet and Miriam Meyerhoff. 1999. The community of practice: Theories and methodologies in language and gender research. *Language in Society* 28: 173–183.

Hymes, Dell. 1974. Foundations in Sociolinguistics, Philadelphia: University of Pennsylvania Press.

Johnstone, Barbara. 1996. *The Linguistic Individual: Self-Expression in Language and Linguistics.* New York: Oxford University Press.

Johnstone, Barbara. 1999. Uses of Southern-sounding speech by contemporary Texas women. *Journal of Sociolinguistics* 3, 4: 505–522.

Johnstone, Barbara. 2000. *Qualitative Methods in Sociolinguistics.* Oxford University Press.

Lanehart, Sonja. 1999. African American vernacular English. In *Handbook of Language and Ethnic Identity,* J. Fishman (ed.). New York: Oxford University Press. 211–225.

Lave, Jean and Etienne Wenger. 1991. *Situated Learning: Legitimate Peripheral Participation.* Cambridge: Cambridge University Press.

LePage, Robert B. and Andree Tabouret-Keller. 1985. *Acts of Identity: Creole-Based Approaches to Language and Ethnicity.* Cambridge: Cambridge University Press.

McCarthy, Teresa and Ofelia Zepeda. 1999. Amerindians. In *Handbook of Language and Ethnic Identity,* J. Fishman (ed.). New York: Oxford University Press. 197–210.

Ochs, Elinor. 1992. Indexing gender. In *Rethinking Context: Language as an Interactive Phenomenon,* A. Duranti and C. Goodwin (eds.). Cambridge: Cambridge University Press. 335–358.

Patrick, Peter. 2002. The speech community. In *The Handbook of Language Variation and Change.* J.K. Chambers, Peter Trudgill, and Natalie Schilling-Estes (eds.). Oxford: Blackwell. 573–597.

Rampton, Ben. 1995. *Crossing: Language and Ethnicity among Adolescents.* Singapore: Longman.

Sacks, Harvey, Emmanuel Schegloff, and Gail Jefferson. 1974. A simplest systematics for the organization of turntaking in conversation. *Language* 50: 696–735.

Sapir, Edward. 1931. Communication. In *Selected Writings of Edward Sapir in Language, Culture and Personality,* David G. Mandelbaum (ed.). Berkeley: University of California Press (1968).

Schilling-Estes, Natalie. 2004. Blurring ethnolinguistic boundaries: The use of "others" varieties in the sociolinguistic interview. Paper presented at LAVIS III, Language variety in the South: Historical and contemporary perspectives. University of Alabama.

Swales, John M. 1990. *Genre Analysis: English in Academic and Research Settings*. NY: Cambridge University Press.

Young, Robert. 2001. The linguistic turn, materialism and race: Toward an aesthetics of crisis. *Callaloo* 24, 1: 334–345.

Wenger, Etienne. 1998. *Communities of practice: Learning, meaning, and identity*. New York: Cambridge University Press.

PART 1

FRAMEWORKS

1

THE DISCURSIVE FRAMING OF PHONOLOGICAL ACTS OF IDENTITY: WELSHNESS THROUGH ENGLISH

Nikolas Coupland

THEORIZING SOCIAL IDENTITY AND LANGUAGE

The last three decades have seen a general shift in social scientific theorizing of identity, from relatively static to more dynamic models, although what these terms mean is itself open to dispute. An early, key voice arguing for this realignment was that of George Herbert Mead in a nascent social psychology (Mead 1932, 1934). Mead argued that the individual's appreciation of social forces in the vicinity of human interaction gave a fuller explanation of what would otherwise have been referred to simply as communicative "behavior." He stressed people's understandings of the social implications of their actions in specific situations, and generally highlighted individuals' agentive capacities in social interaction (discussed in Coupland 2001a). Much later, in anthropology, Frederick Barth's model of ethnicity was rather similarly intended to correct a static, structural-functional understanding of the social world (Barth 1969, 1981). In his historical review of anthropological research on ethnicity, Richard Jenkins argues that this Barthian perspective has come to underpin current conventional wisdom (Jenkins 1997, 12). Barth suggested that we should attend to relationships of cultural differentiation, and that, by focusing on the sorts of boundary work that people do (which of course includes what they do stylistically as part of discursive social action), we can gain an understanding of cultural difference. This is as opposed to the cataloging of trait differences between different ethnic or social groups, which would amount to a static approach to identity. Many recent perspectives on social identity in different disciplines chime with these influential views lobbying for a dynamic perspective. Anthony Giddens argues for seeing identity as a personal project pursued reflexively by people as they navigate through the styles and stages of their lives (Giddens 1991). Theorists in cultural studies have argued vociferously against the assumption that people inhabit unitary identities. Iconic texts in this tradition include Edward Said's treatise on the repressive politics of "Orientalism" (Said 1978). Cultural hybridity and the repressive nature of "essentialising" and "othering" perspectives (Coupland 2000; Riggins 1997) have become normative but not unproblematic assumptions across a wide range. In anthropological linguistics, Richard Bauman and Charles Briggs's theorizing of cultural reproduction as performance, and their concepts of cultural entextualization and decontextualization (Bauman and Briggs 1990), situate the analysis of cultural identity in the domain of discourse. The rather new field of discursive psychology, heavily influenced by Conversation Analysis, extrapolates from Harvey Sacks's insights on social category displays into conversational research on ethnic and other group categorizations, provocatively suggesting that "social identities are for talking" (e.g., Antaki and Widdicombe 1998).

The formative and continuing influence of ethnography on sociolinguistics has guaranteed sociolinguistic engagement with what I am calling the dynamic conception of social identity. This is most obviously the case in Dell Hymes's foundational agenda for an ethnography of communication (e.g., Hymes 1974) and in the interactional tradition of sociolinguistics closely associated with the work of John Gumperz (e.g., Gumperz 1982). All the same, the sociolinguistic study of dialect variation, spearheaded by the remarkable, programmatic research of William Labov into mainly phonological variation and change (e.g., Labov 1972), has tended to downplay the interactional constitution of social identity. This is an important caveat, especially when the "variationist" or "Labovian" or "socio-phonetic" or "secular" tradition of sociolinguistics is held to *be* "sociolinguistics" *tout court* (e.g., by Chambers 1995 and many others). Variationist sociolinguistics has typically taken a static view of social identity, presupposing the integrity of speech communities and working with simple demographic criteria for community membership. Focusing most sharply on the descriptive facts of social distribution and inferable mechanisms of linguistic change, variationist sociolinguistics has been less interested in dialect variation as a locus of social identity work in situated interaction. Many design features of variationist research effectively preclude this perspective, including the preference for relatively large-scale survey techniques, statistical treatments of variation data based on aggregated values of the frequency of dialect variants across speakers, and linear conceptions of dialect "standardness." (I have taken up this issue in more detail elsewhere—see Coupland 2001b for a review.)

What possibilities are there for reconciling the sociolinguistics of language variation with the ever-increasing social scientific trend toward a dynamic view of social identity? In fact there is already a substantial body of sociolinguistic research directed at achieving such an integration. Instances include Howard Giles and his colleagues' research on dialect and "accent" aspects of speech accommodation (e.g., Giles et al. 1991); Allan Bell's theorizing of variation in terms of audience design (originally formulated in Bell 1984); Penelope Eckert's ethnographically based studies of sociolinguistic style as a productive marking process in subcultural groups (Eckert 2000); many of the chapters in the (2002) collection edited by Penelope Eckert and John Rickford on *Style and Sociolinguistic Variation;* and the powerful interdisciplinary critique of sociolinguistic essentialism entailed in Ben Rampton's theoretical and empirical studies of linguistic crossing (e.g., Rampton 1995, 1999). Many other important contributions are omitted from this list. In my own research I have developed a perspective on "dialect style" in the service of managing social personas in interaction (e.g., Coupland 1980, 1988, 2001c). Taken together, these approaches articulate a view of linguistic variation as a dynamic semiotic resource for constructing and managing speakers' social identities and social relationships in ongoing interaction. They stress the strategic nature of sociolinguistic options and uptakes, in some cases formalized as predictive models. They forge a crucial link between accounts of social structure—the architecture of sociocultural differences to which speech features are indexically linked—and social actors' agentive initiatives—what speakers do by way of self-presentation and relationship negotiation.

The essence of this perspective was captured, perhaps more suggestively than elsewhere, in Robert Le Page and Andree Tabouret-Keller's *acts of identity* framework (Le Page and Tabouret-Keller 1985), and this provides my theoretical starting point in this chapter. Viewing language variation as accomplishing acts of identity sits comfortably in the dynamic, constructivist tradition I describe above. In fact, Le Page and Tabouret-Keller (1985, 207) quote Barth (1969) to the effect that "we can assume no simple one-to-one relationship between ethnic units and cultural similarities and differences." This motivates what turns out to be their rather extreme constructivist stance, which is well summarized in a famous dictum:

> the individual creates for himself [*sic*] the patterns of his linguistic behaviour so as to resemble those of the group or groups with which from time to time he wishes to be identified, or so as to be unlike those from whom he wishes to be distinguished. (Le Page and Tabouret-Keller 1985, 181)

This is a view of sociolinguistic projection as a creative dialogic process. They continue: "the speaker is projecting his inner universe, implicitly with the invitation to others to share it . . . and to share his attitudes towards it," reaching out to others who may or may not endorse the cultural validity of what is projected. The summary account seems applicable to a wide variety of sociolinguistic circumstances, not only to new and creole-based communities of the sort Le Page and Tabouret Keller dealt with empirically in their own research, and to monolingual variation as well as to code-choice in multilingual settings. In fact, when I review how I have set out a theoretical agenda for my own work on monolingual dialect style in Wales, I see it mainly as an attempt to illustrate and to specify what the acts of identity framework more generally posits.

What I take to be the core of the approach is to emphasize that some linguistic objects have social *indexicality*—a readable history of sociocultural associations, implications, and therefore "social meanings." Though this is a universally shared assumption in sociolinguistics, an acts of identity approach construes indexical features to be *resources* made available to interactants for certain sorts of identification and relational work in speech encounters. Speakers exercise a degree of control in selecting from a *repertoire* of these resources, in anticipation of and in the service of wanted social outcomes. So we have not only sociolinguistic "behavior" (a term that seems to normalize and neutralize dynamism) or sociolinguistics "variation" (a term that seems to focus analytic interest on linguistic systems rather than on social actors), but *strategic* sociolinguistic *action*. Indexical features are not so much "used" as "deployed," a term that I think echoes Le Page and Tabouret-Keller's term "projection." "Deployment" opens up possibilities of complex ownership relations between speakers and styles, of the sort I deal with below.

My aim is not to review the Le Page and Tabouret-Keller framework in greater detail here, nor to examine its very close relationship to other

established perspectives (those of Howard Giles and Allan Bell in particular). Rather, I want to treat it and them as a body of existing sociolinguistic theory, at least within the camp prioritizing dynamic and
constructivist perspectives. I then want to explore how we might further
refine that general approach by bringing in certain considerations from
discourse analysis, and *frame analysis* in particular (Goffman 1974). So
my starting point is not that the acts of identity perspectives are fully
formed or theoretically exhaustive. In fact I want to argue that there are
several major theoretical gaps and dilemmas that merit further research.
I list several of them in a perfunctory way in the next section, before
dealing in more depth with *one* of them, at the end of my list, as a way of
introducing the two fragments of data I want to focus on in this chapter.

THEORETICAL ISSUES FOR AN
ACTS OF IDENTITY FRAMEWORK

Here are some of the theoretical issues that a dynamic, acts-of-identity-
type orientation to linguistic variation will need to address. I offer the
list of summary points purely as a suggested research agenda for future
consideration, without attempting to resolve them in this discussion.

The Multidimensionality of
Social Identification

The phrase "wanting to identify with" in the Le Page and Tabouret-
Keller quotation is heavily ambiguous—as to ownership and commitment. Projecting a social identity is not the same as feeling or living a
social identity with personal investment in it and full ownership of it—
if identities can in fact be "owned." The subjective/affective/affiliative
dimension easily gets lost in practice-oriented theories of social identity,
just as practice and achievement, and process as a whole, tend to get
lost in both descriptivist and cognitivist approaches. Sociolinguistics
may need to operate with a tripartite model of social identification
through language of the following sort (see Coupland et al. 2003;
Wray et al. 2003):

(1) *Knowledge* of what distinguishes the social group from others,
    and of indexical relationships, where knowledge is presumably a
    prerequisite for engagement with or ownership of an identity.
(2) *Affiliation* to the group's values and distinctiveness, where
    "belonging" can be felt to be "essential," or alternatively aspira-
    tional, routine, irrelevant, etc.

(3) *Practice* what one does to model, symbolize, or enact the culture; how one deploys relevant semiotic material, contextually.

The argument that "language and identity" has to be an interdisciplinary program seems difficult to resist. Different research methods will be needed for addressing different dimensions of identity.

The Scope of Identity Work

The sociolinguistic literature has tended to work with different, over-lapping assumptions of what the "stakes" are in social identification. Sociolinguistic styles, including the monolingual phonological styles that I am concerned with in this chapter, are projected in connection with a diversity of individual or social processes or projects:

(1) *Group work* to assert or project membership in (or not in) various social categories. The categories include ethnicity (e.g., Welshness, Englishness, and their emic subcategories), social class (which is in some ways confounded with ethnicity in Wales), gender groups, specific social networks, and so on (see Giles and Johnson 1981).

(2) *Self work* to reconfigure a speaker's own perceivable personal qualities and traits, for example, to accentuate or deaccentuate attributes of competence, or likeability.

(3) *Relational work* to symbolically manipulate intimacy/distance between people.

Particular theories often treat these dimensions selectively. Yet selves or personas are constructed partly in group terms, as unique constellations of social identities; group boundary work is often done between individuals; the styling of personal/social identity inevitably impacts on how relationships of various sorts are configured. This is why I have previously tried to characterize phonological style-shift, and "dialect style," as the negotiation of "relational selves" (Coupland 2001b).

Establishing Agency

What are the methodological limits to what we can claim or infer from data about the "designing" of identities? Le Page and Tabouret-Keller suggest that, in projection, "by verbalising as he does, [the speaker] is seeking to reinforce his [*sic*] models of the world, and hopes for acts of solidarity from those with whom he wishes to identify" (1985, 181). But evidencing social and communicative goals of this sort is notoriously difficult. Also, communicative goals are multidimensional.

Establishing Outcomes

How do we know that identities are achieved or even "marked"? For whom? What do uptake and change look like?

Categories of Ingroup and Outgroup

The acts of identity notion of "groups we wish to identify with" assumes a situation where a speaker orients to known outgroups. But we also have to address the (probably much more common) case of speakers identifying with *their own* social groups, which they may recognize and model sociolinguistically with varying degrees of detail and precision. Indeed, in each of the two data extracts I consider below, the speakers in question are performing acts of identity to position themselves, in some sense, within their own communities.

Continuous Contact

We tend to approach the study of style, such as dialect styling, with a "first-shot" assumption about speakers and a "no-change" assumption about communities. So, style projection is modeled as the creative deployment, in a fresh context, of established outgroup meanings attaching to features of linguistic styles. In fact, this claim is a cornerstone of Bell's audience design framework, widely debated in Eckert and Rickford (2001)—that stylistic variation is a second-order process, putting to work the social meanings generated by durable correlations between speech styles and speech communities. An example is a speaker shifting to a more "upper class" pronunciation, invoking the values that are associated with an upper-class community. My own arguments about persona management have lived with the same limitation—that we have not as yet made much attempt to model the sociolinguistic processes by which "community speech values" are reproduced or modified. "Speaker's first-shot" and "no change in the community" assumptions work well enough as an account of dialect styling in an extended, "secondary" speech repertoire, for example when a speaker playfully imitates nonlocal voices in humor or in parody (see Coupland 1985). But, relevant to the point made just made above, the semiotic reach of a speaker's stylistic projections are usually a within-community affair.

Cultural Reproduction

Can we isolate speech events or genres that fashion (rather than just reflect) indexical relationships between language and community? Following Bauman and Briggs's line (see also Bauman 1996; Urban 1996), it seems possible to argue that certain classes of communicative events have a special role as sociolinguistic norm enforcers. There are, in a certain sense, "pedagogic" environments for sociolinguistic learning and affirmation. This is to pick up on Max Weber's argument (cf. Jenkins 1997, 10) that ritual, performance events are particularly implicated in sustaining social norms. The important facilitative dimension might well be sociolinguistic reflexivity (Jaworski et al. 2004): events that are strongly reflexive may have a special role to play in cultural reproduction. As a further working hypothesis, we can suggest that the cultural meanings of dialect styles are actively promulgated by a relatively small set of individuals, who, after Giddens, we can call "guardians" of culture (Coupland 2001a; Giddens 1996, 63). This line of theorizing has led me to include an extract of stylized, reflexive performance—a sequence of pantomime talk—in the analysis to follow.

Social Meanings Afforded in Discursive Frames

As this chapter's particular theoretical concern, I hope to show that an acts of identity framework needs to engage systematically with how communicative events are *framed*, and one very obvious point can be made first in this connection. Our enthusiasm to track the functioning of linguistic features or styles in social identification often blinds us to the wider contexts of talk in which they operate. How do we know that, say, phonological variation is the decisive semiotic factor, relative to what people say or do in other dimensions of discourse? This problem is usually discussed in terms of "salience." Social interaction often leaves certain social identities latent, and the linguistic features and styles that might index them remain as unactivated meaning potential. Linguistic and other semiotic features and styles somehow need to be contextually "primed" before sociolinguistic indexing or iconization (cf. Irvine 2001) can occur. To read the identity significance of "dialect in use" or "sociolinguistic variation in discourse," we therefore need to locate speech variables within an integrated discourse analytic perspective (cf. Garrett et al. 1999, 2003).

Interactions like the two I consider below suggest that the identificational value and impact of linguistic features depends on which discursive frame is in place (Goffman 1974). That is, particular discursive frames posit specific *affordances* and *constraints* for interactants at specific moments of their involvement, foregrounding certain types of identity work that can be done at those moments, and either giving relevance or denying relevance to certain categories of linguistic indexicals. I suggest this is so in relation to at least three types or levels of frame:

(1) *The sociocultural framing of relevant communicative events (macro-social frames)*. This refers to the sociolinguistic ecology of particular speech communities. We have to ask what linguistic resources are made available by the sociolinguistic structure of a community, what sociopolitical value systems, perhaps to do with social class or ethnic group membership, do these resources enter into indexically, and what stakes are there to play for in relation to them. At this level, identity work involves speakers positioning themselves or others in relation to prefabricated sociopolitical arrangements in a relevant community.

(2) *The generic framing of communicative events (meso-social frames)*. Generic frames set meaning parameters around talk in relation to what mode or genre of talk, for example, conversation versus set-piece performance, is ongoing and relevant. Identities will be constructed partly in relation to that generic framework, for example, in terms of participant roles. These might confirm or might contradict the identities foregrounded in the wider sociocultural frame of social action, or might supplant them altogether. Participants might find their identity options prefigured or constrained by the generic context, or the genre might edit away identity options that would otherwise apply. The same feature that would mark a sociopolitical identity in the sociocultural frame might carry different resonances in the generic frame. Genres as ways of communicating are typically sustained by particular communities of practice—aggregates of people "who come together around some enterprise" (Eckert 2000, 35). But the normative expectations of practice communities will typically be more local than those of whole sociocultural groups.

(3) *The interpersonal framing of relevant communicative acts (micro-social frames)*. The issue here is how participants dynamically structure the very local business of their talk and position themselves relative to each other in their relational histories, short and long term. Personal and relational identities can be forged and refined linguistically in subtle ways within a consolidated genre and community of practice. A sociolinguistic feature that might otherwise bear, say, a social class or a participant role significance might do personal identity work, styling a speaker as, for example, more or less powerful within a

particular relationship, or might style a speaking dyad as more or less intimate.

I hope to show how each of these broad dimensions of discourse framing needs to be taken into account, by actors and analysts, in understanding the identity work done through phonological variation in each of two data extracts I now turn to. The implication of this argument is that claims about the apparently inherent social meanings of phonological features, such as those made in language attitudes research, have to be treated with some caution. Useful generalizations have certainly been made about, for example, the social meanings of "standard and non-standard accents" in terms of perceived competence and social attractiveness (Garrett et al. 2003; Giles and Powesland 1975; see Eckert 2001, 122 for comments on the meanings of some central phonological variables in U.S. English). I do not at all mean to imply that this work should not be carried forward, and in fact it will be the best way of filling out the social meanings made available by different sociocultural frames. But the dynamics of social identity work will also need to take account of more local contextualizing factors.

The two data extracts are from very different social contexts—one from the world of popular theatrical performance, the other from a workplace setting. The extracts share the geographical context of English language being used in south Wales in the United Kingdom. The theatrical performance in question is a Christmas pantomime, performed and videotaped in a south Wales Valleys theater in front of a live audience. The second extract is from a travel agency in the center of Cardiff, the capital city of Wales, involving a group of female assistants who develop small talk amongst themselves around their more formal professional talk with clients and holiday operators. The sociolinguistic ecosystems in south Wales—Valleys and Cardiff— where these two very different speech events take place do have their unique qualities, socially and linguistically. However, these are not crucial to the line of analysis I develop below, so I make only a few comments about them. My motive in choosing the two extracts is to see whether engaging with different levels of discursive framing can make it possible to read identity work across radically different types and contexts of talk with some degree of theoretical coherence.

The Pantomime Dame

The first extract is from a Christmas pantomime, *Aladdin*, performed in late December 2001 at a theater in a small town in the south Wales

Valleys. The show was toured around other theaters across south Wales, although its cultural roots are firmly "Valleys." The Valleys have a long tradition of heavy industry, especially coal-mining and the production of iron and steel, but suffered drastic economic decline through the middle and late decades of the twentieth century. Left-wing political radicalism in Britain historically found its most influential leaders in the Valleys, which retain this political feel as well as a structural poverty that is slow to ease. The pantomime is produced by and stars a well-known local radio and television performer, Owen Money, who plays the character Wishy Washy. Owen Money is an apologist for English-language Valleys speech and cultural values in his radio, TV, and live shows in Wales, and he is prominent in the Valleys community in other ways too; he is, for example, director of Merthyr Tydfil football club.

The British phenomenon of pantomime is not easy to explain to people unfamiliar with the genre. It is a generally low budget, low-culture, burlesque form of music, comedy and drama, with a live orchestra. The form is generally holding its popularity. "Pantos" run at very many theaters through England and Wales over the months of November, December, January, and February, being thought of as Christmas entertainment but not thematically linked to Christmas itself. Pantomimes are often said to be entertainment for children, although family groups make up most audiences. Each pantomime theme is a variation on one of a small number of traditional narratives, with roots in folk tales. Each theme tends to mingle ethnic and temporal dimensions with abandon. This performance of *Aladdin*, like the animated Disney films of that title, builds its plot around an Arabian Nights magic lamp and a magic genie. But the performance also uses stage sets including "Old Peking," and the Wishy Washy character's name refers to his menial job in a Chinese laundry. Pantomime plots always involve magic, intrigue, royalty, peasantry, and a love-quest. Typically, a noble and honorable prince, conventionally played by a female, dressed in a tunic and high boots, falls in love with a beautiful girl from a poor family. The girl has either large, ugly, vain sisters or a large, ugly, vain mother, referred to as a Dame and often named The Widow Twanky (these females are conventionally played by males). Characters are starkly drawn and heavily stylized. Young love triumphs and royalist grandeur is subverted, which is not an unpopular theme outside of theater in contemporary Wales. The semiotic constitution of pantomime is bricolage, intermixing light popular songs and comedy routines, exorbitant colors and costumes, and with vernacular, self-consciously "common" values set against regal pomp and transparently evil figureheads. The interactional format involves a good deal of audience participation and ingroup humor. Hackneyed and formulaic plots are interspersed with disrespectful humor on topics of local or contemporary interest. Conventional teases appeal to children, who have to shout warnings to the heroine princess, for example, when an evil emperor approaches, or to help the audience's friend (in this case Wishy Washy) to develop his quest (e.g., to find the magic lamp).

The extract below is the pantomime Dame/Widow Twanky's first entrance, close to the beginning of the show after the opening song performed by the full cast and live orchestra. The Dame's entrance is a tone-setting moment for the whole pantomime. She is the mother of Aladdin, the nominal hero, and she returns regularly through the pantomime, mainly to add the most burlesque dimension of humor on the periphery of the plot. Next to Wishy Washy, she is affectively "closest" of all the characters to the audience. Pompous, vain, and mildly salacious, she is nevertheless funny and warm-hearted. Her transparent personal deficiencies leave her open to be liked, despite them.

The most striking socio-phonetic contrast in the extract is between the Dame's aspirationally posh, mock-Received Pronunciation (RP) voice at the opening of the extract, and the broad vernacular Valleys Welsh English voice that she otherwise uses. The principal variable speech features that carry this contrast are listed in table 1.1, where the first-listed variant in square brackets in each case is the "standard," RP-like variant. Italicized lexical forms are items appearing in the transcript.

Table 1.1 Phonological variables for south Wales Valleys English

(ou)—[əʊ], [oↄ], [oː]	(*hello, home, nose; w*idow has only the diphthong options)
(ei)—[ei], [eː]	(*name, later,* but not *hey, day, anyway,* which again have only the diphthongal variant)
(ʌ)—[ʌ], [ɚ]	(*brothers, lovely, bunch*)
(ai)—[ai], [ɚi]	(*died, time, find, bye*)
(iw)—[juː], [jɪw], [ɪw]	(*you,* where the "local" variant has a prominent first element of the glide, contrasting with the RP-type glide to prominent /uː/)
(ɔ)—[ɔ], [ↄɚ]	(*poor*)
(a)—[æ], [a]	(*grans, grandads, back, Twanky, man, had, Lanky, manky, hanky, stand, Aladdin*)
(h)—[h], [Ø]	(*hello, home, hey, husband, hanky, he*)
(ng)—[ŋ], [n]	(*gossiping*)

Extract 1: The pantomime dame
(enters waving, to music "There is nothing like a dame")
1 hello everyone
2 (Au: hello)
3 hello boys and girls
4 (Au: hello)
5 hello mums and dads
6 (Au: hello)
7 grans and grandads brothers and sisters aunts and uncles
8 and all you lovely people back home ooh hoo
9 hey (.) now I've met (.) all of you
10 it's time for *you* to meet (drum roll) *all* (.) *of* (.) (cymbal) *me*
11 (Au: small laugh)
12 and there's a lot of *me* (.) to *meet* (chuckles)
13 now my name is (.) the *Wid*ow *T-wan*ky
14 and do know what (.) I've been a widow now (.) for *twen*ty-*five years* (sobs)
15 (Au: o:h)
16 yes (.) ever since my poor *hus*band died
17 oh what a *man* he was (.) he was *gor*geous he was
18 do you know (.) he was the *tall*est man (.) in *all* of Peking
19 and he always *had* (.) a *runny nose* (chuckles)
20 hey (.) do you know what we *called* him?
21 "*Lan*ky *Twan*ky with a *Man*ky *Han*ky"
22 (Au: laugh)
23 hey (.) and guess what (.) I've still got his manky hanky to this *ve*ry *day* look look at that ugh
24 (Au: o:h laughs)
25 hey (to orchestra) look *af*ter that for me will you?
26 you look like a *bunch* of *snobs*
27 (Au: laugh)
28 anyway (.) I can't stand around here *goss*iping all day
29 *I* have got a *laun*dry to run
30 ooh (.) and I've got to find my *two* naughty boys (.) A*ladd*in (.) and *Wish*y *Wash*y
31 so (.) I'll see you lot later *on* is *it?*

32   (Au: ye:s)

33   (to camera) I'll see *you* later on (.) bye for now (.) tarra (.) bye bye

    (leaves waving, to music "It's a rich man's world")

Lines 1–8 show centralized onset of (ou) in all three tokens of *hello* and in *home*, contrasting with monophthongal [o:], which occurs later in the word *nose*. We also have fully audible [h] in all cases in these opening lines. Together, these features carry the symbolism of "posh" as the Extract opens, apparently outgrouping the Dame relative to the Valleys community in which the performance is geographically and ideologically situated. Aitch-less *hey* at line 9 and the schwa realization of the first syllable of *brothers* (in place of the wedge vowel) mark a strong shift from a conservative English RP voice into Valleys vernacular. The RP voice resonates most strongly at line 8 in the utterance *all you lovely people back home*, where the first two and last two words have significant RP and nonlocal tokens. The abrupt stylistic shift indexes a cracked or unsustainable posh self-presentation, a chink in the Dame's dialectal armour of "posh," which is thereby confirmed to be as suspect as her dress-sense. The wider semiotic dimension here is fundamentally to do with authenticity and inauthenticity.

After line 8, all tokens of (iw) have the Valleys local form, including *you* in line 10, said with contrastive stress. The Dame's self-introduction in line 13 pronounces word *name* with the vernacular form [e:], although *Widow T-wanky* (with a prolonged /w/ glide), when she mentions her name, reverts to the conservative RP centralized form. This achieves a neat splitting of personas, between the introducing voice and the introduced voice, phonologically pointing up the Dame's inauthenticity. The sequence setting up the *manky hanky* wordplay (meaning "disgusting handkerchief") is performed in a fully formed local vernacular. All three vowels in the stressed syllables of *poor husband died* (line 16) are local Valleys variants. Similarly, aitchless *he* on the three occasions in line 17 and monophthongal *nose* in line 19 are prominent.

The Dame's vernacular style is realized lexico-grammatically too. We have reduplicative *he was* at line 17, the word *manky* (meaning "disgusting"), the invariant tag *is it?* at line 31 (which, more usually in its negative form *isn't it?*, is a strong stereotype of Welsh English), and colloquial *tarra* for "good bye" at line 34. Discursively too, the mock formality of the opening salutation and self-introduction is counterpointed (and confirmed to have been mock) by later stances. The

Dame's feigned grief at being widowed is subverted by the joke at the husband's expense and by references to the Dame's large bosom and hips (see lines 10 and 12). The disrespectful wordplay, *bunch of snobs* (*snobs* evoking "snot" or nose effluent, visually rendered by the bright green stain on the handkerchief), addressed to the orchestra builds an allegiance against the conservative persona she feigns early on, and so on. How do these stylistic selections impinge on our readings of identity in the extract? Pantomimes, and performance events generally, provide data of an entirely irrelevant sort, according to canonical sociolinguistics. The social identification potential of dialect is generally assumed to be activated in the real language of real speech communities, where authentic members imbibe social values during socialization and proceed to recycle these values indexically in their vernacular speech throughout their lives. The variationist project has partly been to find methodological means of accessing the untrammelled vernacular in all of its purity and regularity, and Widow Twanky's dialect performance therefore stands well outside of the canon. In fact, the interface between variationist sociolinguistics and authenticity is an interesting and productive one, and one that is beginning to be critically explored (Bucholtz 2003; Coupland 2003). Without opening up such issues in detail here, we can nevertheless consider ways in which staged and stylized dialect performances *can* become interpretable as identity work, provided that analysis respects the various levels of discursive framing I introduced earlier. Let's first consider the generic framing of the Widow Twanky sequence, which is where the most obvious contextualizing constraints are operative in this case.

The Generic Frame

Pantomime is theatrical performance, and in a sense self-consciously "bad" performance, at least in relation to wider norms of theater. Characters in pantomime engender the usual theatrical complexities of ownership—whose voices are these? whose identities? is everything feigned? is it all "just for entertainment"? As I noted above, the genre is thoroughly conventional and ritualized; it is burlesque and extravagant in its visual, rhetorical, and vocal forms. Its "talk" is self-reflexive as to character, plot, and humor; Widow Twanky is self-parodic, knowingly inauthentic, highly stylized. A thin and close-to-the-surface plot, overdrawn characters, and visible performances are endemic in the genre (see the Dame's stark mentioning of plot elements at lines 29–31). The staging of the event is itself generally transparent, for example, in performers' frequent references to the co-present audience (see the Dame's talk to the audience at lines 1–8 and 29–34, her talk to members of the orchestra, etc.). In the extract she purports that she has kept her husband's handkerchief for twenty-five years, but both calls it a *manky hanky* and gives us a token affective response to its disgustingness (*ugh* at line 24).

So this particular genre frame rules out several identity options. The actor's own personal identity is clearly irrelevant. The formulaicity of the plot effaces identity work that might otherwise have related to the Dame's "personality" in the story frame. Although there is gender-layering, gender identity is obscured by the conventional transgendering that the genre requires, and so on. But the extract nevertheless offers us several relationships that we can and must make sense of, under the constraints of a performance frame. One is the relationship between the Dame's two personas, the socio-phonetically indexed posh and local Valleys identities. Although the genre prevents us from reading these identities as relevant to either the actor or the character of the Dame herself, the very conventionality of the genre invites us to see meaning in the "posh"/"local" contrast in some wider sociocultural frame. Overdrawn images, like pictorial cartoons, have the characteristic of wide semiotic applicability, precisely because the stylization that produces their "broad-brush" and "bright color" features obliterates particular reference. Then there is the relationship between the Dame and the audience, who are directly addressed in the text. The extract shows the Dame switching reference and address between the (fictional) *Aladdin* plot world and the (real) Valleys theater world and its *boys, girls, mums, dads,* and so on. The Dame exists in both domains, but not as a "straight" inhabitant of either. Nevertheless, she does draw the audience into particular alignments with some of her espoused stances, with identity implications for audiences. The Widow Twanky's acts of identity are certainly indirect and conflicted projections. But, if only by her studious and extravagant efforts to deauthenticate herself according to non-panto norms, she opens up other possibilities in the sociocultural and interpersonal frames.

The Sociocultural Frame

If we look at the wider ideological climate in which this pantomime performance operates in the south Wales Valleys, there are clear group referents for the Dame's two stereotyped personas, posh and vernacular Valleys. Valleys vernacular English lacks a clear prestige standard within its own territorial boundaries; RP is not a significant stylistic resource for predominantly working-class Valleys people, even though they are of course aware of RP as a powerful outgroup status variety—in England and to a lesser extent in the capital city of Cardiff, some twenty miles to the south. Then, unlike west and especially northwest Wales, and in a different way again Cardiff, the Valleys have only a limited dialog with the Welsh language revival. Our earlier research (e.g., Garrett et al. 2003) shows Valleys English to be a variety that is heavily stigmatized without having the "compensating" attributions of social attractiveness or, despite its working-class heritage, a high degree of "real Welshness."

In this ecosystem, the Dame's inauthentic posh voice is definitely non-Valleys and probably non-Welsh. The dialect personas she is projecting play out a familiar ideological conflict with powerful ethnic and social-class resonances. The discourse of the extract is organized around this contrast, with ideational ("content") and dialect meanings interwoven into it. The Dame's initial, showy, public persona is done in the RP outgroup voice. As we have seen, the first element of private, apparently self-deprecatory reference (line 9) is where the RP voice begins to crumble. The mock desolation at the death of her husband (line 16) is done in the intimate and parochial voice. But her persona is in fact resilient, in that she gives us evidence that her grief is inauthentic. It is fabricated to tell a silly joke at his expense, also at her own expense. So the sequence indirectly projects significant cultural authenticities, about a cultural group that is resilient despite its low prestige and poverty. In the sociocultural frame, the Dame's slipping mask of pretentious, conservative RP English is a jocular form of subversion—of (English) social-class hegemony. This is a wholly ludic context for political satire, but the Dame does offer a counter-identity for a Valleys audience—as "one of us," someone who was only feigning and failing to aspire to a higher class. Hostility to English snobbery is even more clearly signaled across the other characters in the pantomime—in an RP-like evil villain; also in the consistently Valleys vernacular speaking Owen Money in the role of Wishy Washy, the "children's friend."

The Interpersonal Frame

Audience members are fully ratified as vocal participants and coperformers in pantomime. In the interpersonal frame, Widow Twanky's shift into an exuberant vernacular Valleys style, allied with her textual references to the orchestra members seated between her and the audience (in their formal suits) as *a bunch of snobs*, draws the audience into specific anti-snobbish dialect-indexed values. There are several moments in the extract when the Dame invites specific responses from the audience. The first is the exchange of greetings, performed in the inauthentic posh voice. These are sequentially integrated turns where the audience does respond audibly, although without any obviously strong integrative affect on the audience's part. Similarly at line 15, when the audience delivers a formulaic *o:h* in response to the Dame's predictable lament about being a widow (she is after all by name "the Widow Twanky"). Affective integration happens most obviously at lines 22, 25, and 28. The first two of these are when the audience is suitably disgusted by the *manky hanky*—a glowingly (green) vernacular icon— and the second is when the audience aligns with the Dame's *bunch of snobs* insult. The interaction creates a space for joint participation and fills it with a vernacular Valleys style, aligned against "posh." A reflexive and stylized public performance in this way reproduces elements of a vernacular culture premised on "authenticity from below."

Travel Agency Assistants

The second extract is a retranscription of some data from a Cardiff travel agency that I first worked on 25 years ago (see Coupland 1980, 1984). The study focuses on a set of assistants working in a city center office, and on one assistant in particular, Sue. I return to the data here partly to assess the gap between my early and current responses to it, and partly because it is in its own terms remarkably rich data for the analysis of dialect style. The phonological variables potentially in play in the assistants' speech as Cardiffians overlap considerably with those listed earlier for Valleys English in the pantomime data. See table 1.2, which repeats many items from table 1.1, identifying the lexical items in which they are potentially operative in the second extract. Table 1.2 then adds variable features that have more specific applicability to Cardiff English.

As the extract opens, Sue is trying to connect on the telephone to a coach tour operator, *Rhondda Travel*. The extract then allows us to follow two concurrent conversations. One is Sue's telephone conversation, where we don't have access to the other party's voice. The other conversation is among the three travel agency assistants, Sue, Marie, and Liz, about buying charcoal, then about eating lunchtime sandwiches. We hear this less formal conversation only partially because of overlapping speech and because the recording microphone is positioned closest to Sue's service position in the office. Sue's talk on the telephone is represented in italics in the extract, to help distinguish the two separate conversational flows. All three women have similar Cardiff vernaculars in what Labov calls their "less careful" speech.

Table 1.2 Phonological variables for Cardiff English

(ou)—[əʊ], [oᴗ], [oː]	(*oh, go, charcoal, hello, hold, don't, so, OK, going, though; know* has only the diphthong options)
(ei)—[ei], [eː]	(*great, take*; but not *today, Friday, pay, anyway, pay, say,* which again can have only diphthongs)
(ʌ)—[ʌ], [ɚ]	(*come, stuff, rubbish*)
(ai)—[ai], [ɚi]	(*I, I've, Friday*)
(iw)—[juː], [jɪw], [ɪw]	(*you*)
(a)—[æ], [a]	(*can, have, Travel, dad, Dallas, Blacks, camping, that, had*)
(h)—[h], [Ø]	(*hello, Hourmont, held, have, hold, had*)
(ng)—[ŋ], [n]	(*shopping, going, talking, camping, starving, going, anything*)
But also:	
(aː)—[ɑː], [aː], [æː]	(*are, charcoal, barbecue, starving*)
(intervocalic t)—[t], [ʈ], [r]	(*but I'm, but I, about Evans*)
(intervocalic r)—[J], [r]	(*where are, they're all*)

Extract 2: Travel agency assistants

```
 1 Sue:        come on Rhondda Travel where are you?
 2                                                    [
 3 ?Marie:                            hm hm hm
 4             (4.0)
 5 Liz:        o:h I got to go shopping where d'you think I can get
             charcoal from?
 6 Sue:        (0.5) I don't really know
 7                                        [
 8 Marie:                        is today Wednesday?
 9 Sue:        yeah
10             [
11 Liz:        Marie (.) if you're going out (.) can you just see if you
12           can see any charcoal anywhere if you're just walk- walking
             around the shops (( ))
13 Sue:       (on the telephone) hello (high pitch) can I have Rhondda
14           Travel please?
15 Marie:     ((            I'm only going           laughs))
16 Liz:        oh (laughs) (high pitch) where you going then?
17 Marie:     I'm going to the solicitors
18 Liz:        oh (laughs) my dad's been up there ((he ought to    ))
```

```
19                                    [
20 Sue:                           hello it's Hourmont
21      Travel here (.) um was I talking to you about Evans (.)
22         [
23 Marie:   ((                    ))
24                                       [
25 Liz:                           barbecue
26 Sue:    to Dallas? well the problem is I've held an option on them
           for you (.)
27                                       [
28 Marie:   ((                         ))
29 Liz:    will they?
30 Sue:    but I can't book them in full cos you have to take full
           payment (1.0)
31         [
32 Marie:   ((                               ))
33 Liz:    do they? I've never seen it
34                        [
35 Sue:                      you see so they'll hold them for me now
           until Friday
36                                             [
37 Marie:                                       ((
38         (laughs)        ))
39          [
40 Liz:       (laughs) they don't sell things like that
41 Marie:   (1.0) course they do
42 Sue:    well I've booked them and they're all alright (.)
43              [
44 Liz:        where would I get it from?
45 Sue:    but I can't give them ticket numbers until they pay
46                        [
47 Marie:                   ((              ))
48                                       [
49 Liz:                           charcoal
50 Sue:    (breathy voice) OK?
```

51 Marie: ((Blacks   ))
52 Liz: yeah *camp*ing stuff innit yeah
53 Marie: ((and Woolies))
54 *Sue:*   *mm (.) alright*
55   [
56 Liz: (( )) reckoned Woolies as well but I don't
57   think so (1.0) I'll just go down to Blacks
58   [
59 *Sue:*   *that'll be great (1.0) we'll let you know if*
60   *you can-o:h Friday morning (.) yeah that's OK the*
              option's till Friday anyway (.)
61 (other client conversations in the background)
62 *Sue:*   *OK then fine (1.0) OK then (.) bye (.) Sue (1.0)*
              (breathy) OK? bye
63 Marie: (faint) is anyone *else* (( )) starving?
64 Sue: well I *was* going to have one but I'm not going to now
65 Marie: well *have* one don't pay any attention to what *I* say
66   [ ]
67 Sue: no
68 Marie: I talk a load of rubbish
69   [
70 Sue: I'd rather you know no you *know*
71   about them don't you
72 Marie: no I *don't* I don't know *any*thing
73 Sue: that's *all* I've had to eat *then* though

Sue is minimally involved in the charcoal conversation early on, at lines 6 and 9. She comes back into the three-way conversation after hanging up the phone at the end of line 62. My main analytic interest is in the transition achieved between lines 63 and 65, as Sue rejoins the triadic conversation to talk about lunch. Only some of the above-listed phonological variables show variation in the extract. In general, the assistants do not use centralized-onset for (ou) in *go*, or wedge in *come*, or the close RP-type variant of /a/ in *can*, all of which would be marked as posh in Cardiff; *great* and *take* show up only in Sue's telephone conversation where RP-like variants occur. On the other hand, (iw) is never RP-like [ju:] in *you*, being [jɪw] throughout.

As audible variation within the extract, Sue has markedly more open onset to (ai) in *Friday* (line 35) than in all the first-person pronouns (*I*) at the end of the extract (lines 64–73). In fact there is a powerful clustering of vernacular variants of the consonantal variables in Sue's speech starting at line 64. In *have* (64) and *had* (73) (h) is [Ø]; *going to* is ['gənə] (64); *about them* is [ə'barəm] (71); *don't you* is ['do:nɪw] (71). In line 1, Sue has fronted [a:] in *are*, before she speaks to the Rhondda Travel representative. Similarly, Liz's *camping stuff innit* at line 52 contrasts starkly with Sue's "careful" speech in the same time slot but in a different conversation.

In the original analysis, I quantified the distribution of "standard" and "nonstandard" variants of some of these variables over much longer stretches of data, in order to demonstrate that mean values for several variables showed systematic covariation across different contexts of speaking—such as Sue speaking "more standardly" on the telephone than off the telephone, when talking about work-related topics as opposed to nonwork topics, and to clients than to her coassistants, and then that she differentiated in a rather precise way, on a quantitative basis, in her talk to different social classes of client. How does that interpretation look now? Though this sort of generalization still seems worth demonstrating, the original analysis does impute a direct semiotic value to phonological variants (such as stop versus flapped intervocalic [t], or [h] versus [Ø]) in carrying social meanings such as "careful" versus "spontaneous," or "middle-class" versus "working-class," and a richer analysis seems warranted. My earlier classification of "contextual types," deriving from Hymes's taxonomy of speech event "components," seems to both overspecify and underspecify how Sue's talk is contextualized. As an alternative conceptualization, let's try to invoke the three-way framing schema once again.

The Sociocultural Frame

The most powerful linguistic-ideological contrast in the Cardiff community relates to social class, as in the classical Labovian urban paradigm. Cardiff, as a large, socially diverse, long-anglicized city, displays the sort of English-language sociolinguistic stratification by class that we see in most major British urban sites. This contextual factor loads up the sociolinguistic variables that are most sensitive to class in Cardiff (but arguably much less so in other parts of Wales). The relevant phonological features include (h), (ng), and the "high/low articulation" variables such as consonant cluster reduction, whose more elided forms are stigmatized as "common" or "slovenly" ways of speaking, as is the case with (intervocalic t). Talk to nonfamiliars in Cardiff, such as the tour operator Sue is dealing with on the telephone in the second extract, is amenable to social class inferencing. Her identity work on the phone is very plausibly class-work, and she may be seeking a more middle-class persona of the sort that tends to gain status in public and especially workplace discourse in Cardiff and other cities. On the other hand, several other factors impinge, which I come to below.

However, still following the social-class theme, it doesn't seem right to say that Sue's identity is salient for its working-class meanings when she is talking about her sandwiches and her dieting, later in the second extract. Being of a social class is neutralized once the frame shifts from public to private discourse, where class is a shared ingroup value, although Sue's being in some ways "powerful" or "powerless" at personal and relational levels *is* relevant. Also, we can't be sure that the class-work is done through phonological indexicality, or solely by this means. Notice how Sue's telephone conversation ideationally invokes commercial power practices. In her own words, Sue has *held an option on* a booking clients for the tour operator who has to *take full payment* before the deal can proceed. Compare this with the "walking round the shops to try and buy charcoal" theme of the competing conversation, or Sue's own powerlessness in the face of a depressing diet at the end of the extract (see later). Class as control *is* relevant in the public projection on the telephone, and class semiosis through dialect constitutes part of Sue's identity in her professional mode of discourse.

The Generic Frame

In terms of genre, however, there are clear transitions between professional talk and everyday-life-world talk in the extract. Overlaid on the social-class reading of Sue's talk, the genre structure positions her as abruptly moving out of the role of professional representative at the end of line 62. She does give her personal name while operating in the professional frame—*Sue*, at line 62—but she does this in that minimalistic form of person reference that is conventional in telephone service encounters. She is the voice of this specific travel agency, Hourmont Travel, and she and other participants may feel that there should be some resonance between her vocal style and a smoothly, competently functioning travel agency. Notice the build-up of professional jargon through Sue's telephone talk. Also the vivid disjunction between Sue's rhetorically abrasive and Cardiff vernacular *come on Rhondda Travel where are you?* (with close front /aː/ in *are*) at line 1 and her concerned, solicitous demeanor as the telephone conversation closes. The genre frame facilitates identity readings in terms of professional versus personal roles as relevant social meanings for Sue's talk.

The Interpersonal Frame

Sue's talk between lines 64 and 73 is not only nonpublic discourse and nonprofessional discourse; it is personally intimate discourse. Its deals with what was a rich topic domain in the travel agency over the many weeks of my recording there—eating and dieting. This is a theme in which the three assistants, and Sue in particular, invest heavily in emotional terms. There is a regular relational politics around dieting among the three assistants, affecting moves to eat lunch at all, and certainly decisions about the timing of *when* sandwiches are eaten. Sue's *I was going to have one but I'm not going to now* at line 64 raises delicate issues. "Having one" here means eating a sandwich before the due lunchtime hour, when it would have become more legitimate to eat, according to the assistants' dieting pact. At line 63, Marie has transgressed by asking if anyone is *starving*, when it's taken for granted that the others, and especially Sue, are self-consciously holding back from eating their sandwiches. Disclosing her eating regime to her coassistants, so that they know what she eats and when (*I'd rather you know . . . about them*, lines 70–71), is a strategy Sue uses to help her to resist early eating.

The sandwiches exchanges invoke issues of entitlement, trust, blame, and potential praise—a moral agenda—in an intimate relationship between the assistants. What part could speech style and "dialect" have in this relational work? One semiotic principle at work at this point in the talk is implicitness. Contrasting sharply with the on-the-surface explicitness about professional procedures in the telephone talk, Sue drops into a way of speaking, triggered by Marie's question about being *starving*, where the dieting agenda, its components, its participant roles, and its pressures are all thoroughly known to the group. Lexico-grammatically, "having one" is sparse. So, discursively, is the coherence link between Sue's saying she isn't going to "have one" and Marie's response that Sue shouldn't pay attention to what she says. The offence and Marie's recognition of it are explicated by the assistants' relational history.

In heavily implicit talk, it is perhaps unsurprising that phonological processes also shift toward elision and economy, and this is what we see in the phonetic description of Sue's final utterances in the extract. But there is also a personal standing or status semiotic dimension in play. Sue is very audibly depressed at having been forced to confront her dieting regime. Perhaps she thinks she is a failure, or at least in need of Marie and Liz's policing of her diet. Her identity work in the interpersonal frame is to mark this "incompetence," and the dialect semiosis does contribute to achieving this. What is made relevant in the interactional frame is neither "lower-class" nor "non-professional status"; it is low personal control. We might gloss the dialect style as "under-performance," which is also marked in reduced amplitude and flatter pitch range.

Conclusion

An acts of identity framework seems very apposite as a general orientation to phonological variation of the sort that surfaces in the two very different instances we have considered. In each case, a speaker can reasonably be said to be projecting social identities, projecting personas that are at least in part fashioned on the basis of indexical relationships between phonological forms and stereotyped social roles. Le Page and Tabouret-Keller's rubric, suggesting that speakers creatively project identities "so as to resemble those of the group or groups with which from time to time [they] wish to be identified, or so as to be unlike those from whom [they] wish to be distinguished" is a rough gloss of the processes I have tried to describe in the pantomime data and the travel agency setting data.

In some ways, however, Le Page and Tabouret-Keller's summary of social processes seems too open. The expression "from time to time" implies a degree of latitude and a degree of opportunism that the two data extracts belie. The sociocultural contexts of the particular speech events certainly act as constraints on what social meanings are available to be constructed and inferred from the two sorts of phonological performance. If we feel that Widow Twanky's playful projections of English posh and Valleys vernacular might not "work" outside of Wales or outside of the Valleys towns, this is to suggest that the social meanings she deploys are in some important way *afforded by* the sociolinguistic structure of the local community. We would also need to be circumspect about the reference "groups" that Le Page and Tabouret-Keller invoke. Certainly posh and vernacular styles have group-level associations, of the sort I have described. But in the generic and interpersonal frames, phonological style forges associations more with communicative roles and traits of personality than with social groups as such. It might be possible to argue that discourse roles and personalities are themselves "group-linked" social phenomena, but that line of argument unduly weakens the importance of genre and selfhood as foci for identity work. Acts of identity need not be restricted to alignments with social groups, even though this is what models such as Allan Bell's audience design model have focused on centrally.

Then there is the question of "identifying" being modeled as a process of "resembling" or "being distinguished from." It has become commonplace to view sociolinguistic style-shifting as the putting on and taking off of social identities, as if speakers were regularly able in some sense to "pass off" from moment to moment as members of different groups, meaning that their overall social identities are "hybrid." The emphasis on *performance* in the two extracts we have considered calls this view into question. Most obviously in the pantomime instance, it would be reckless to claim that Widow Twanky is variably attempting actually to pass as posh or as a Valleys vernacular speaker. As a character, and certainly as an actor behind the mask, she/he is surely attempting to do neither of these things. A performance frame undermines direct claims to the inhabitation and ownership of social identities, and it is more suitable to talk in terms of *reference* and *mention* than in terms of *ownership* and (in the usual sociolinguistic sense) *use*. This is what Rampton's work on sociolinguistic crossing has made clear, particularly in those moments of dialect stylization that he deals with. The issue that arises is how the acts of identity perspective interprets "resembling," when "resembling and passing as" is a radically different process of social identification from "resembling without passing." The latter points up social differences whereas the former seeks to obscure them. Generalizations about social trends are difficult to support in this area. But it may prove to be the case that, in the socially and generically complex and increasingly reflexive social circumstances that late modernity offers us, identity work may become less and less a matter of multiple ownerships and transferred allegiances. It may become more and more a matter of navigating individual paths through complex semiotic structures, and of salvaging fragments of personal identity from the various social consonances and dissonances we and others are able to set up discursively.

The local contextualization of identity work will become more important, and it is the acts of identity framework's *under*-specification of local sociolinguistic processes that I have attempted to address in this chapter. Although the two extracts we have dealt with show some definite overlaps in the meanings of the phonological resources that are available to speakers of English in south Wales, it would ultimately be misleading to seek overarching generalizations about dialect and identity on the basis of a link between, say, "dialect standardness" and, say, "social class identity." I have suggested that the various contextual frames work both as constraints and as affordances. They are constraints in that they close off specific potential meanings (e.g., the gender identity of the pantomime Dame or Sue's social identity when she agonizes about her diet). They are affordances in that they open up specific meaning clusters at particular discursive moments (e.g., extrapolating to an ethnic Welsh / English conflict from the Dame's playful style-shift or Sue's symbolizing of her low personal control in relation to her eating regime). This suggests that the "salience" of a sociolinguistic feature or style is not only, or perhaps not even principally, related to its perceptual prominence or its place in a phonological system or its frequency of use, although these may be relevant factors. Rather, salience is a quality of a sociolinguistic feature that is potentiated by its use in a particular social and discursive frame, where it becomes available to do specific sorts of identity work and not others.

A debate about structure versus agency (or about inherent versus contingent identities) has surfaced and resurfaced in the sociolinguistics of style. Yet a framing perspective shows this debate to have been based on a false dichotomy. For all the local construction work we can evidence in how speakers manage their own and others' social identities, sociolinguistic styles do have some recurrent social values, which themselves therefore have some ontological status. In consequence, we do not have to be committed to a social constructionism of radical ephemerality. We do not have to believe that social identities, as the discursive psychologists have it, are purely opportunistic and "for talking," nor that our social and personal identities are refashioned anew on each occasion of talk. In fact, a framing perspective forces us to track how specific identity potentials, established and remade in a community's structure, *are or are not* operationalized in specific contexts and moments of talk.

Nor do we have to give up on people's subjective investment in social identities. The framing complexities I am pointing to do not leave speakers as necessarily "hybrid" beings, chameleon-like and identificationally puny social creatures who change their sociolinguistic coloration from one moment to the next, which was one implication of Le Page and Tabouret-Keller's dictum. The concept of social hybridity through language is arguably a loose generalization resulting from a failure to track how social identities are constructed in all their contextual complexity. An adequate sociolinguistic perspective on discursive social meaning—which it has tended to label reductively as the study of "style"—needs to attend to both the regularities of sociolinguistic structure and such regularities as we can establish in how local contexts of talk motivate and facilitate social identity work. These then need to be treated as the backdrop against which interpersonal dynamics work, with or against social norms, as the frames of social interaction are built and broken. The sociolinguistics of "style" might in fact be defined as analyzing how indexical linguistic resources are deployed and interpreted in the light of what particular contextual frames afford and preclude as realizable and relevant social meanings.

References

Antaki, C. and S. Widdicombe (eds.). 1998. *Identities in Talk*. London: Sage.

Barth, F. (ed.). 1969. *Ethnic Groups and Boundaries: The Social Organisation of Culture Difference*. Oslo: Universietsforlaget.

Barth, F. 1981. *Process and Form in Social Life: Collected Essays of Frederik Barth*, Vol. 1. London: Routledge and Kegan Paul.

Bauman, R. 1996. Transformations of the word in the production of Mexican festival drama. In *Natural histories of discourse*, M. Silverstein and G. Urban (eds.). Chicago and London: University of Chicago Press. 301–327.

Bauman, R. and C. Briggs, C. 1990. Poetics and performance as critical perspectives on language and social life. *Annual Review of Anthropology* 19: 59–88.

Bell, A. 1984. Language style as audience design. *Language in Society* 13: 145–204.

Bucholtz, M. 2003. Sociolinguistic nostalgia and the authentication of experience. *Journal of Sociolinguistics* 7: 398–416.

Chambers, J.K. 1995. *Sociolinguistic Theory: Linguistic Variation and Its Social Significance*. Oxford: Blackwell.

Coupland, N. 1980. Style-shifting in a Cardiff work setting. *Language in Society* 9: 1–12.

Coupland, N. 1984. Accommodation at work: Some phonological data and their implications. *International Journal of the Sociology of Language* 46: 49–70.

Coupland, N. 1985. "Hark, hark the lark": Social motivations for phonological style-shifting. *Language and Communication* 5: 153–172.

Coupland, N. 1988. *Dialect in Use: Sociolinguistic Variation in Cardiff English*. Cardiff: University of Wales Press.

Coupland, N. 2000. "Other" representation. In *Handbook of Pragmatics: Instalment 2000*, Jef Verschueren, Jan-Ola Ostman, Jan Blommaert, and Chris Bulcaen (eds.). Amsterdam and Philadelphia: John Benjamins Publishing Co.

Coupland, N. 2001a. Introduction: Sociolinguistic theory and social theory. In *Sociolinguistics and Social Theory*, N. Coupland, S. Sarangi, and C.N. Candlin (eds.). London: Longman/ Pearson Education. 1–26.

Coupland, N. 2001b. Language, situation and the relational self: Theorising dialect style in sociolinguistics. In *Style and Sociolinguistic Variation*, P. Eckert and J. Rickford (eds.). Cambridge and New York: Cambridge University Press. 185–210.

Coupland, N. 2001c. Dialect stylisation in radio talk. *Language in Society* 30, 345–375.

Coupland, N. 2003. Sociolinguistic authenticities. *Journal of Sociolinguistics* 7, 417– 431.

Coupland, N., H. Bishop, and P. Garrett. 2003. Home truths: globalisation and the iconisation of Welsh in a Welsh–American newspaper. *Journal of Multilingual and Multicultural Development* 24, 3: 153–177.

Eckert, P. 2000. *Linguistic Variation as Social Practice*. Maldon, MA. and Oxford: Blackwell Publishers.

Eckert, P. 2001. Style and social meaning. In *Style and Sociolinguistic Variation*, P. Eckert and J. Rickford (eds.). Cambridge: Cambridge University Press. 119–126.

Eckert, P. and J. Rickford (eds.). 2002. *Style and Sociolinguistic Variation*. Cambridge: Cambridge University Press.

Garrett, P., N. Coupland, and A. Williams. 1999. Evaluating dialect in discourse: Teachers' and teenagers' responses to young English speakers in Wales. *Language in Society* 28: 321–354.

Garrett, P., N. Coupland, and A. Williams. 2003. *Researching Language Attitudes: Social Meanings of Dialect, Ethnicity and Performance*. Cardiff: University of Wales Press.

Giddens, A. 1991. *Modernity and Self-Identity: Self and Society in the Late Modern Age*. Cambridge: Polity Press (in association with Basil Blackwell).

Giddens, A. 1996. Living in a post-traditional society. In *Reflexive Modernization: Politics, Tradition and Aesthetics in the Modern Social Order*, U. Beck, A. Giddens and S. Lash (eds.). Cambridge: Polity Press. 56–109.

Giles, H. and P. Johnson. 1981. The role of language in ethnic group relations. In *Intergroup Behaviour*, J.C. Turner and H. Giles (eds.). Oxford: Blackwell. 199–243.

Giles, H. and P. Powesland. 1975. *Speech Style and Social Evaluation*. London: Academic Press.

Giles, H., J. Coupland, and N. Coupland (eds.). 1991. *Contexts of Accommodation: Developments in Applied Sociolinguistics*. Cambridge: Cambridge University Press.

Goffman, E. 1974. *Frame Analysis*. Harmondsworth: Penguin.

Gumperz, J.J. 1982. *Discourse Strategies*. Cambridge: Cambridge University Press.

Hymes, D. 1974. *Foundations in Sociolinguistics: An Ethnographic Approach*. Philadelphia: University of Pennsylvania Press.

Irvine, J.T. 2001. "Style" as distinctiveness: The culture and ideology of linguistic differentiation. In *Style and Sociolinguistic Variation*, P. Eckert and J. Rickford (eds.). Cambridge and New York: Cambridge University Press. 21–43.

Jaworski, A, N. Coupland, and D. Galasinski (eds.). 2004. *Metalanguage: Social and Ideological Perspectives*. Berlin: Mouton de Gruyter.

Jenkins, R. 1997. *Rethinking Ethnicity: Arguments and Explorations*. London: Sage.

Labov, W. 1972. *Sociolinguistic Patterns*. Philadelphia: Pennsylvania University Press.

Le Page, R.B. and A. Tabouret-Keller. 1985. *Acts of Identity: Creole-Based Approaches to Language and Ethnicity*. Cambridge: Cambridge University Press.

Mead, G.H. 1932. *Philosophy of the Present*. LaSalle, IL: Open Court.

Mead, G.H. 1934. *Mind, Self and Society*. Chicago: Chicago University Press.

Rampton, B. 1995. *Crossing*. London: Longman.

Rampton, B. (ed.). 1999. Styling the other. Special issue of *Journal of Sociolinguistics* 3, 4.

Riggins, S.H. (ed.). 1997. *The Language and Politics of Exclusion: Others in Discourse*. Thousand Oaks: Sage.

Said, E. 1978. *Orientalism*. London: Routledge and Kegan Paul.

Urban, G. 1996. Entextualization, replication and power. In *Natural Histories of Discourse*, M. Silverstein and G. Urban (eds.). Chicago: University of Chicago Press. 21–44.

Wray, A., N. Coupland, H. Bishop, and B. Evans. 2003. Singing in Welsh, becoming Welsh: "Turfing" a "grass roots" identity. *Language Awareness* 12, 1: 49–71.

2

A Sociolinguistics of "Double-Consciousness": English and Ethnicity in the Black Experience

Alamin A. Mazrui

Introduction

This essay seeks to explore aspects of the relationships between English and "ethnicity" in the global African experience, as intergroup and intragroup processes. The topic itself has been inspired, in part, by the notion of "double consciousness," a concept usually associated with W.E.B. DuBois to describe that peculiar tendency of the black person

> . . . of always looking at one's self through the eyes of others, of measuring one's soul by the tape of a world that looks on in amused contempt and pity. One ever feels his two-ness—an American, a Negro; two souls, two thoughts, two unreconciled strivings, two warring ideals in one dark body, whose dogged strength alone keeps it from being torn asunder. (DuBois 1997, 38).

Though DuBois was specifically describing the African American condition of being, the concept itself is equally relevant to the rest of the black world: In the latter case, however, the two-ness would involve the schism between "Negroness" and "humanness" within a context in which the terms of that humanness are defined by the "racial" (i.e., European) other.

When it is a product of the extent to which the black person has "interiorized the racial stereotypes" of the hegemonic other—hegemony in the sense employed by Antonio Gramsci—DuBois's two-ness constitutes only one strand of Frantz Fanon's idea of the two-ness of the black person. The other strand describes the black person's sense of being relative to other black people. Together, the two strands constitute a two-dimensional black persona. In Fanon's words:

> The Black person has two dimensions. One with his fellows, the other with the white man. A Negro behaves differently with a white man and with another Negro. That this self-division is a direct result of colonialist subjugation is beyond question. (Fanon 1967, 17)

Applying this idea to the French language from the point of view of a psychologist, Fanon observes how the behavior of the black person of the French Antilles contrasts with his/her behavior when in the company of other black people from the "Francophone" world. The only time the black person assumes an "independent self" in relation to the French language is when the interlocutor is "foreign" to the language (Fanon 1967, 36), unable to judge his/her linguistic "Frenchness."

This essay draws partly from this Fanonian view of the two-dimensionality of the black person, exploring its linguistic implications with specific regard to the English language in the black world. In the process, however, I shall discuss other relevant issues that fall outside the ambit of this theoretical paradigm.

In addition, there are three caveats that I would like to make in connection with Fanon's formulation. First, each of the dimensions of Fanon's concept must itself be seen as multidimensional, assuming various shades depending on such factors as nationality, ethnicity, class, religion, and gender. A middle-class British-trained African male, for example, is likely to have a different sense of his linguistic Englishness vis-à-vis a middle-class Briton than a middle-class American. Similarly, the place of English as a medium of communication between fellow black people takes somewhat different configurations of meaning depending on who is talking to whom, when, how, and where. The symbolism of English when a Yoruba elite in Nigeria is in the company of his/her fellow ethnic compatriots from the rural areas, for example, may be quite different from the symbolism of the language when it is employed in conversation with fellow Yoruba elite, and different still when used with the elite from other ethnic groups. Although my essay will not explore all these multiple levels of the relational universe of English, it is important to bear in mind this wider sociolinguistic complexity.

Second, Fanon was right that the origins of this two-dimensional character of the black person can be traced back to the fact of colonial domination. But the significance of the concept is by no means limited to the colonial experience. Some scholars have indeed extended Fanon's understanding of the link between the colonizer and the colonized to noncolonial systems altogether—to relations of patriarchy, for example. This two-dimensionality, then, would probably hold true under many conditions of domination by an "other" that has succeeded in establishing its ideological hegemony. To this extent, Africa's neocolonial reality would be as valid a context for the exploration of the two-dimensional quality of the Black person as the colonial condition.

Third, Fanon framed the black–white strand of the two-dimensional orientation only in terms of "submissive dependency" on the linguistic terms of reference established by the racial "other." But there are instances in which the rejection of those same terms can constitute a kind of "aggressive dependency" on them in a way that also betrays the black person's dimension vis-à-vis the white person. When the Nation of Islam proclaimed, in its formative years, that the "white person is a devil," for example, it was in fact accepting a Eurocentric axiom of racial determinism that had been employed to inferiorize the black person.

Bearing in mind these three qualifications, then, we can now proceed to consider how this two-dimensionality has manifested itself as a linguistic articulation of the black person in the "Anglophone" world and some of the other English-related issues of ethnicity that are manifest in global Africa.

ENGLISH AS A UNIFYING FORCE

At the heart of the controversial Oakland School Board's decision on "Ebonics," perhaps, was a proclamation of an independent black linguistic identity vis-à-vis the European other within the North American context. And this history of the politics of identity is probably related to the "racial boundaries of the English language," a condition that can be appreciated best by comparing it with, say, the Arabic language. Any person who speaks Arabic as a first language could, in principle, claim Arab ethnic affiliation. This contrasts with English, which does not admit into its "Anglo ethnic fold" people who are not genetically European. As a result, African Americans could not associate themselves with the dominant Anglo-American identity simply by virtue of being "native" speakers of English. Had the American lingua franca been Arabic instead of English, on the other hand, the entire African American population today could have been ethnically Arab (Mazrui and Mazrui 1998, 30–31).

It is true, of course, that this "ecumenical" quality of English has sometimes been the source of its strength, especially in situations of strong ethnolinguistic nationalism. Different people around the world may feel comfortable to make the language "their own" partly because, in doing so, they do not have to assume the identity of the other. The assimilative tendency of Arabic, on the other hand, may trigger the fear of imperialism. Protective of his Dinka ethnic culture and identity, Kelueljang criticizes his cousin in the following verses:

> My cousin Mohamed
> Thinks he's very clever
> With pride
> He says he's an African who speaks
> Arabic language,
> Because he's no mother tongue!
>
> Among the Arabs
> My cousin becomes a militant Arab—
> A black Arab
> Who rejects the definition of race
> By pigment of one's skin.
>
> He says,
> If an African speaks Arabic language
> He's an Arab!
> If an African is culturally Arabized He's an Arab! (Quoted in
> Chinweizu 1988, 35)

In the racial climate of the United Sates, however, it is the ethnic exclusiveness of English that African American nationalism has tended to react against, leading to an African American quest for alternative sources of ethnolinguistic identity. And the reaffirmation of the autonomy and uniqueness of Ebonics became part of this identitarian exercise. This condition is what may have led Molefi Asante to claim that the "prototypical language of African Americans has been named *Ebonics* to distinguish it from English" (1987, 35).

The sentiment in favor of a peculiarly black version of the English language, however, is by no means limited to the American scene. It is also found elsewhere in the "Anglophone" regions of the black world, even in Africa where there is a strong presence of local languages tied to specific ethnic identifications. In South Africa, for example, there has emerged a whole movement of "People's English," a form that is deemed to be different from "international English." As one advocate of People's English comments:

> To interpret *People's English* as a dialect of international English would do the movement a gross injustice; *People's English* is not only a language, it is a struggle to appropriate English in the interests of democracy in South Africa. Thus the naming of *People's English* is a political act because it represents a challenge to the current status of English in South Africa in which control of the language, access to the language, and teaching of the language are entrenched within apartheid structures. (Pierce 1995, 108)

This South African effort, no doubt, is one with which the renowned African writer Chinua Achebe is in agreement, in part, when he suggests that the African writer "should aim at fashioning out an English which is at once universal and able to carry his own experiences . . . But it will have to be a new English, still in full communion with its ancestral home but altered to suit its new African surroundings" (Achebe 1965, 29–30).

It is true, of course, that this black nationalism that claims a peculiarly black English (in all its diversity) is itself triggered, at times, by the seeming attempt of "native" white speakers of English to be possessive about the language and monopolistic about setting its standards of correctness. When a certain Englishman once complained about the degeneration of English in Kenya, for example, back came the following reply from Meghani, a non-British Kenyan:

It is not at all wisdom on the part of a tiny English population in this wide world to claim that English, as presented and pronounced by Americans, Canadians, Africans, Indians, and the people of Madras State, is not English. It may not be Queen's English, but then what? Has the Englishman the sole right to decide upon the form and style of a universal language?

Meghani then goes on to argue that, "Strictly speaking, English cannot be called 'English' at all, since it is a universal language belonging to all. It is difficult to understand why it is still known under that horrible name; it should have had another name" (*East African Standard* (Nairobi) February 15, 1965). Meghani thus sought legitimacy for particularistic varieties of English—including black ones—by appealing to its universality.

ENGLISH AND PAN-AFRICANISM

But these attempts, in different regions of the black world, to inscribe an English or Englishes that bear the imprint of Africanity tend to mask the central role of "mainstream (British and American) Englishes" in the politics of black identity. There is a sense in which global Africa can be described as a synthesis of the racial heritage of Africa and the linguistic heritage of Europe. Racial Africanity has provided the bonds of shared identity; the languages of Europe have often provided the network of shared communication between black people. In the final analysis, then, the black Englishes that exist or are presumed to exist as markers of black identity, often capitulate to approximations of American and British "standard" varieties of the language as a way of fostering linkages between black people toward a pan-African identity.

The place of English as a language with the potential to unify black people has received special attention with regard to the African continent, particularly because of the scope of its linguistic diversity. As early as the 1880s, the pioneer pan-Africanist Edward W. Blyden, for example, regarded the multiplicity of "tribal languages" in Africa as divisive and believed that this linguistic gulf could be bridged best by English than by any other European language partly because English itself was a product of a multicultural heritage. In the words of Blyden:

> English is, undoubtedly, the most suitable of the European languages for bridging over the numerous gulfs between the tribes caused by the great diversity of languages and dialects among them. It is a composite language, not the product of any one people. It is made up of contributions by Celts, Danes, Normans, Saxons, Greeks and Romans, gathering to itself elements . . . from the Ganges to the Atlantic. (Blyden 1888, 243–244)

As African Americans were reaching out to be reunited with their ancestral land, the unity of the continent itself was seen to be at stake. The English language provided a possible bridge.

It is, of course, rather curious that Blyden favored English over Arabic as the language of continental pan-Africanism. After all, he was a minister who repeatedly praised the role of Islam and the Arabic language in Africa. Blyden very much desired to launch an Arabic program at Liberia College where he was already a professor of "classics." He celebrated Arabic as a language that had contributed to the cultural growth of Africa as "already some of the vernaculars have been enriched by expressions from Arabic" (Lynch 1971, 270). Yet, when the chips were down, and in spite of his Islamophilia and Arabophilia, he supported English precisely because, in his mind, English was a synthesis of various ethnic languages to which no individual people could lay absolute claim.

As history would have it, however, English became a potential tool of communication not only between Africans across the ethnic divide, but also between people of African descent across the seas. And because of the racial politics of their historical time and place, the English-speaking African Americans came to assume a particularly central place in the leadership of transcontinental (political) pan-Africanism, especially in what was emerging to be the Anglophone black world.

In the postcolonial period, the consolidation of English in Africa, ironically, has been aided in part by forces of ethnic nationalism. Nationalism is usually regarded as a political ideology that is concerned about the value of its own culture and with protecting it against "external" encroachments. But, in the context of power politics of the African nation-state, the "out groups" are often perceived to be, not the "non-African other," but members of other African ethnic constituencies. Under the circumstances, the quest for a national language has often tended to favor English (and other European languages) because giving the language of any one ethnic group some official status over the others is seen as potentially hegemonic. When Nigeria once considered having Hausa as the national language, for example, Chief Anthony Enahoro is reported to have said in the Nigerian parliament: "As one who comes from a minority tribe, I deplore the continuing evidence in this country that people wish to impose their customs, their languages, and even more, their way of life upon the smaller tribes" (quoted by Schwarz 1965, 41). Chief Enahoro was a strong advocate of English as the country's national and official language partly because of the fear of internal ethnic domination.

Upon reflection, then, if Blyden wished for English to become the trans-ethnic language of Africa, his dream has been moving closer and closer to becoming a reality. You may forge pan-Africanism in North Africa and rely exclusively on the Arabic language. You may attempt a pan-African union in East Africa and rely mainly on the Swahili language. But, for the time being, neither the Organization of African Unity nor the newly formed African Union has been conceivable without resort to English and French languages.

For some scholars, the value of English goes well beyond its bridge-building potential across different black "tribes" of the world. It extends its power to the construction of an African consciousness itself. According to this school of thought, the very sense of being African as a collective experience would have been impossible without the instrumentality of the English language. Ken Saro-Wiwa, the Nigerian writer who was executed in 1995 in the course of struggle for the ethnic rights of his own Ogoni people, was particularly assertive of this view:

> With regard to English I have heard it said that those who write in it should adopt a domesticated "African" variety of it. I myself have experimented with the three varieties of English spoken and written in Nigeria: pidgin, "rotten," and standard . . . That which carries best and which is most popular is Standard English, expressed simply and lucidly . . . And so I remain a convinced practitioner and consumer of African literature in English. I am content that this language has made me a better African in the sense that it enables me to know more about [fellow Africans from] Somalia, Kenya, Malawi, and South Africa than I would otherwise have known. (Saro-Wiwa 1992, 157)

A similar sentiment was expressed by Leopold Sedar Senghor, the first president of Senegal, who, in spite of his strong Francophilia, claimed that English has "been one of the favorite instruments of the New Negro, who has used it to express his identity, his *Negritude*, his very consciousness of the African heritage" (Senghor 1975, 85).

In a poem entitled "The Meaning of Africa," the Sierra Leonean poet Davidson Abioseh Nicol defined "Africa" in the following manner:

> You are not a country, Africa
> You are a concept
> Fashioned in our minds, each to each,
> To hide our separate fears
> To dream our separate dreams

And what Saro-Wiwa and Senghor are suggesting is that the English language was an indispensable stimulus to the very birth of that concept, painful as the birth process itself was.

ENGLISH AND AFROCENTRICITY

A related dimension of black consciousness is more epistemological in its claims and has come to be known as Afrocentricty. But what is Afrocentricity and how does it relate to pan-Africanism? We define Afrocentricity as a view of the world that puts Africa at the center of global concerns and idealizes its role in human affairs. It puts great emphasis on the agency of black people in shaping not only their own history, but the history of the world at large, ascribing to people of

African descent a greater role in the construction of human civilization than has been recognized. In the final analysis, Afrocentricity seeks to restore the pride and confidence of black people in their own African heritage.

Pan-Africanism, on the other hand, is a doctrine or movement that believes in the common destiny of African peoples and seeks to unite them politically, economically, and culturally. Whereas Afrocentricity regards Africa as a cultural complex in the widest sense of the word and is inspired by the idiom of black dignity, Pan-Africanism sees the continent primarily as a political entity and its idiom draws heavily on the spirit of solidarity. Of course, neither of these ideologies is monolithic.

Within the United States, Afrocentricity seems caught between the instrumental value of English and the symbolic value of indigenous African languages. The instrumental value can include both a collective scale (of fostering community bonds, for example) and individual scale (of serving the communicational needs of individual users). The symbolic value, on the other hand, relates more to concerns of collective identity, consciousness, and heritage.

The symbolic use of African languages within Afrocentricity coincides with a quasi-Whorfian position. Afrocentrists draw on culture-specific words—those with complex and language-specific meanings (as in the often quoted example of multiple terms for snow in Eskimo), and cultural key words, the highly salient and deeply cultural-laden words (e.g., "honor" in Arab society as compared to "freedom" in American society) (Dirven and Vespoor 1998, 145)—from Africa's linguistic pool in their attempts to center Africa as the modal point of their ideology. The instrumental side that is pegged to English, on the other hand, is predicated on a "functionalist" view of language (Hawkins 1997). The concern here is not with how language influences cognition, but with how language itself is (re)structured in terms of the functions to which it is put. Racial assumptions and biases and exclusionary ideologies are not inherent in language, but are reflected, perpetuated, and naturalized in the way language is used. Within this framework, then, Afrocentrists see the English language as an instrument by which to inscribe the black experience within which black people are grounded in a racially divided society entrapped in a hegemonic ideology that is decidedly Eurocentric.

And how do Afrocentrists seek to resolve the seeming tension between their Whorfian and functionalist positions? They actually do not. But, in general, they seem to regard language as operating on two planes: One that is particularistic, reflecting a heritage of black people in Africa and its Diaspora, shaped by their historical experience over the centuries; the other, more plural (or universal?)—malleable and potentially amenable to a multiplicity of accommodations (though often through a process of struggle and contestation). African languages are mobilized toward the particularistic mission whereas English is deemed subject to "multiculturalization."

Many nationalists within the continent of Africa tend to advocate for the replacement of European languages inherited from the colonial tradition by African ones. In the forefront of this campaign has been the Kenyan writer Ngugi wa Thiong'o, who has repeatedly argued that "the domination of a people's language by languages of the colonizing nations was crucial to the domination of the mental universe of the colonized" (1986, 16). The process of radical decolonization proposed by Ngugi, therefore, involves a rejection of English, the subsequent refusal to submit to the worldview supposedly embedded within it, and the recentering of African languages in the intellectual life of African peoples.

For Afrocentrists in the West, however, the range of linguistic alternatives to Eurocentrism is much more circumscribed. With English as their first and often the only language, African Americans cannot easily exercise the kind of total linguistic shift advocated by African nationalists. The linguistic challenge confronting the Afrocentrist, then, has been how to articulate counterhegemonic and anti-Eurocentric discourses in a language of "internal" domination.

In an effort to meet this challenge, one path that has been pursued by Afrocentrists has been the "deracialization" of English. This process has sometimes involved attempts to inscribe new meanings (e.g., in the word "black") or to create new concepts (e.g., *kwanzaa*) in the language so as to make it more compatible with the dignity and experiences of black people. Molefi Asante provides a list of examples of English words today, which, in his opinion, "must either be redefined or eliminated" altogether because they belong to the kind of language that "can disrupt the thought of good solid brothers and sisters" (1989, 46–47).

Asante is, of course, quite cognizant of the fact that Eurocentrism in language transcends lexical semantics or meanings inscribed in individual words and phrases. It exists, rather, in the entirety of its symbolic constitution. Beyond the level of specific words that are "monoethnic," we are told, "there are substantive influences upon language (a sort of Whorfian twist) that make our communicative habits sterile. The writers who have argued that English is our enemy have argued convincingly on the basis of '*black*ball,' '*black*mail,' '*black* Friday,' etc; but they have not argued thoroughly in terms of the total architecton of society" (Asante 1987, 55). The Afrocentric challenge, then, is seen as one of subverting the entire symbolic generation of "mono-ethnic" (i.e., Eurocentric) meanings in an otherwise plural world.

The deracialization of English among Afrocentrists has also taken the form of particularizing what had hitherto been portrayed as universal. When we make inference to "classical music"—a phrase invariably taken to refer to the compositions of people such as Beethoven, Bach, and Mozart—Afrocentrists insist on knowing whose classical music we are talking about. Terms like "discovery," "modern languages," and many others are similarly subjected to this relativist reinterpretation, which allocates meanings to their specific cultural-experiential contexts. As Tejumola Olaniyan aptly put it:

> Instead of one world, one norm, and many deviants, Afrocentric cultural nationalism authorizes several worlds with several norms. The universalist claim of Europe is shown to be a repression of Otherness in the name of the Same. "Culture," as the West erects it, is hence subverted to "culture," "Truth" to "truth," "Reason" to "reason," "Drama" to "drama." This is the fundamental ethicopolitical point of departure of the Afrocentric cultural nationalist discourse, an empowerment of a grossly tendentiously misrepresented group to speak for and represent itself . . . (1995, 35)

What is involved, ultimately, in this attempted recodification in the terrain of language and discourse is a struggle over who has the right to define, the right to name.

Some Afrocentrists also believe that there is a certain Eurocentric structuring of thought in the construction of knowledge that is promoted partly through the English language. They associate with English certain conceptual tendencies including, for example, dichotomization (e.g., reason versus emotion or mind versus body), objectification and abstractification (where a concept is isolated from its context, its place and time, and rendered linguistically as an abstract). These features, it is argued, are in contradiction to the human essence and reality—seen to be integral to Afrocentric thought—and their end result is the fortification of a Eurocentric ideology with all its conceptual trappings (Ani 1994, 104–108).

All in all, then, in embracing English as their own, Afrocentric thinkers have refused to accept its idiom passively and uncritically. And, sometimes, they have risen to the challenge of constructing new and imaginative metaphors and meanings. They have aimed to follow in the tradition of Nat Turner and Henry Highland Garnet, two important figures in African American protest history, who are said to have stood "against the tide of Europeanization in their discourse even though the representational language was American English," the language of their oppressors (Asante 1987, 126).

Even as they seek to transform it, however, English has continued to serve as the main medium of an Afrocentric counter-discourse. Much of the theorizing about Afrocentricity and the formulation of models based on it has been done in English. And it is with the facilitating role of the English language that Afrocentricity gets communicated to black people both within the United States and beyond. It is in this sense of articulation and communication of ideas that we have ventured to suggest that Afrocentricity is dependant on the instrumental value of the English language.

But in the attempt to affirm an African identity, to devise maxims based on that identity, and to construct a symbolic bridge between the African Diaspora and African cultures, Afrocentrists have often had to turn to African languages. Yoruba, for example, has come to feature quite prominently in libation rituals in many an Afrocentric gathering. Kariamu Welsh-Asante (1993) partly draws from the Shona language of Zimbabwe to define the conceptual parameters of an Afrocentric aesthetics. And in spite of the fact that Alexis Kagame's work (1956) has been discredited by some African philosophers (e.g., Masolo 1994, 84–102), his propositions of an "African worldview" based on the categories of his native language, Kinyarwanda, have continued to exercise a strong influence on Afrocentric thinkers in the United States. In the words of Dona Richards, Kagame has made it possible for Afrocentric intellectuals "to express African conceptions in African terms" (1990, 223).

From the entire corpus of African languages, however, it is Kiswahili that has been Afrocentricity's most productive source of symbolic enrichment. Indeed, according to Maulana Karenga, African Americans have the same kind of claim to Kiswahili as Jews, for example, have to Hebrew. "Swahili is no more frivolous or irrelevant to black people than Hebrew or Armenian is to Jews and Armenians who were not born in Israel or Armenia and will never go there" (Karenga 1978, 15). Kiswahili is the language of the most serious challenge to Christmas to have emerged in the African Diaspora. Inspired by African harvest ceremonies as markers of temporal cycles, an entire idiom drawn mainly from Kiswahili has come into existence to designate *Kwanzaa*, the African American end of the year festival, and its *Nguzo Saba* or seven pillars of wisdom. These include *Umoja* (Unity), *Kujichagulia* (Self-determination), *Ujima* (Collective responsibility),

Nia (Intention), *Kuumba* (Creativity), *Ujamaa* (Socialism), and *Imani* (Faith). Every December, hundreds of thousands, if not millions, of African Americans celebrate *Kwanzaa* in the name of Mother Africa.

English as a Diversifying Force

But English has not only been key to black unity and black consciousness; it has also stimulated new identities within the black world. Though it has linked black populations from various continents, English has worked in tandem with other European languages to reconstruct the black world into Anglophone, Francophone, and Lusophone blocs. Within the Anglophone domain, there has been the divide between African American Vernacular English, Caribbean English, British black English, and several varieties of African English. George Bernard Shaw once said that England and America are two countries divided by a common language, English. Here we have black folk scattered in three continents who are also divided by that same common language of European origin.

Between Americo-Liberians and Afro-Saxons

Within Africa, the earliest divisive effect of English came with the establishment of the colony of Americo-Liberians, the African American repatriates that came to settle in the West African country of Liberia. Americo-Liberians became a distinct ethnic group in their own right—demarcated away from indigenous blacks by differences in lifestyle and by the English language as a standard of "civilized" speech. The linguistic attitudes of the time were well captured by the pioneer pan-Africanist, Alexander Crummell, who regarded African languages as lacking in "clear ideas of Justice, Human Rights, and Governmental Order, which are so prominent and manifest in civilized countries" (Crummell 1969, 20). English, on the other hand, was seen to possess the opposite credentials. In Crummell's deterministic words:

> . . . the English language is characteristically the language of freedom. I know that there is a sense in which this love of liberty is inwrought in the very fibre and substance of the body and blood of all people; but the flame burns dimly in some races; it is a fitful fire in some others; and in many inferior people it is a flickering light of a dying candle. But in

the English races it is an ardent, healthy, vital, irrepressible flame; and withal normal and orderly in its development. (Crummell 1969, 23)

He saw Africans exiled in slavery to the "New World" as inheritors of "at least this one item of compensation, namely, the possession of the Anglo-Saxon tongue" (1969, 9). And he wished for the rest of the black race this same divine providence given to African Americans. He regarded the linguistic Anglicization of Africa, with Americo-Liberians as its pioneers, as a necessary step toward Africa's civilization.

Once English became established in Liberia, however, it remained the only African country for a while that owed its English to America. Other Africans on the continent who were exposed to the English language at all, were so exposed through their encounter with British colonialism. With post–Cold War globalization, however, there is evidence of increasing American influence on the English varieties spoken in Africa, even though the rate of this linguistic change may vary from place to place.

But if Liberia has its own ethnolinguistic class of Americo-Liberians, we see in much of the rest of Anglophone Africa the emergence of a new transnational "tribe" of Afro-Saxons. These are, in Ali Mazrui's definition, Africans who speak English as a first language, often as a direct result of interethnic marriages, especially at the level of the elite.

As the father and mother come from different linguistic groups, they resort to English as the language of the home. English thus becomes the mother tongue of their children, with a clear ascendancy over the indigenous languages of both the father and the mother. (Mazrui 1975, 11)

In South Africa, the offspring of white and black parentage are a distinct ethnic group called "Colored." Will Afro-Saxons, the offspring of mixed ethnic unions, one day become conscious of themselves as a group independent of the ethnic affiliations of their parents? There is some impressionistic evidence that an "Afro-Saxon" consciousness is indeed in the making.

The irony of Afro-Saxons, of course, is that while they are a group alienated from many of their ethnic and national compatriots, they are the most trans-ethnic, transnational and transcontinental Africans in linguistic affiliation. Across the Atlantic, African Americans once led the pan-African movement partly because of their facility with the English language. In his discussion on the origins of pan-Africanism, George Padmore also tells us about the role of English in forging a trans-ethnic national consciousness among the Creoles of Sierra Leone (1956, 39). Will Afro-Saxons now be in the forefront of the pan-African movement, on a subcontinental, continental, or transcontinental scale by virtue of the primacy of English in their lives? The answer is obviously in the womb of time.

ENGLISH, APARTHEID, AND ITS AFTERMATH

The ethnic dynamics of the English language have a different manifestation altogether in the Republic of South Africa, partly because of the character of the country's white constituency. Of all the African states, of course, South Africa has always had the largest white population, estimated at five million. But this population is by no means monolithic: Within it are differences that are maintained by marriage patterns, residential zones, ethnic-based commercial networks, and so forth.

Until the 1990s, the great divide between black and white in South Africa was indeed "racial." But the great cultural divide between white and white was, in fact, linguistic. The white "tribes" of South Africa were the Afrikaans-speaking Afrikaners, on one side, and English-speaking Europeans, on the other. Language had "tribalized" the white population of South Africa.

In time, however, this linguistic division between the white "tribes" of South Africa also came to have its own impact on the black population of the country. More and more black South Africans felt that if they had to choose between English and Afrikaans, the former was of greater pan-African relevance. Two Germanic languages had widely differing implications. Afrikaans was a language of racial claustrophobia; English was a language of pan-African communication. The Soweto riots of 1976, precipitated in part by the forced use of Afrikaans as a medium of education in African schools, were part of that linguistic dialectic.

With the end of political Apartheid in South Africa, the English language has made the clearest gains. Although South Africa has declared eleven official languages (theoretically reducing English to one-eleventh of the official status), in reality the new policy only demotes Afrikaans, the historical rival of English in the country. English has continued to enjoy the allegiance of black people, almost throughout the country, as the primary medium of official communication.

This seeming consolidation of English in post-Apartheid South Africa has inspired a new wave of Afrikaner nationalism, triggered by the fear that their language and identity would be compromised by the new linguistic dispensation. But the development has also stimulated an uneasy alliance between a section of "Coloreds" and the Afrikaners. These cross-ethnic allegiances are particularly pronounced in the arena of party politics, of which ethnic group is allied to which political party. In the words of Michael Chege, "Fears that their language and identity will be swallowed by the new South Africa undergirds much Afrikaner resistance to ANC rule. This also accounts for the National Party's popularity among many of the part-Dutch 'coloreds' in the Cape—the so-called brown Africans, whose primary language is Afrikaans" (Chege 1997, 79).

ENGLISH, BETWEEN ACCESS AND ACCENT

But, as suggested earlier, in the multiplicity of functions that English has played in Africa, one has been to plant new seeds of diversity between its inhabitants. One of the most prominent English features of black diversity in Africa is, of course, that of "pronunciation." Ethnically marked varieties of English are legion in many parts of the continent and usually, to the experienced ear, it is not difficult to tell, from the English accent alone, who is a member of which ethnic group.

As much as Africans regularly make fun of each other's ethnically marked accents of English, however, attitudes prevail in some quarters that members of "our" ethnic group speak better English than our "other" ethnic compatriots—with "better" judged from a foreign standard of propriety, from an imagined approximation to British Standard English. At times this linguistic attitude is accompanied by an ethnocentric belief that, consistent with the English language yardstick, "our" ethnic group is somehow more culturally sophisticated than "other" ethnic groups. This is the same tendency that Fanon observed in the "Negro of the Antilles" in his relationship with "natives" from Francophone Africa, or in the attitude of Martinicians toward Negroes of Goudeloupe (Fanon 1967, 25–27).

But how real are these competitive claims of members of different African ethnic groups about their command of English? A leading African scholar, Ali Mazrui, once suggested that members of Afro-Islamic ethnic groups (such as the Hausa of Nigeria and the Swahili of Kenya) "have been both among those who have been relatively suspicious of the English language as a factor in cultural transformation and among those who have shown an aptitude for speaking it well" (Mazrui 1975, 54). With regard to Nigeria, specifically, he points out that one of the ironies of the English language in that country is that southerners (such as the Yoruba and the Igbo) have better *access* to the English language than northerners (e.g., the Hausa-Fulani), but supposedly northerners have better *accents* for the English language than southerners (personal communication, July 4, 2002).

Afro-Muslim suspicions of English and the relatively easier Afro-Christian access to it can be traced back to the interplay among the language, education, and Christian missions in the colonial period. There was even a time when English proficiency was often associated with an Afro-Christian background. But it is said that when Africans from Afro-Islamic ethnic groups "have finally capitulated to the pull of the English language as a medium of intellectual modernity, they have been among the better speakers of the language" on the continent (Mazrui 1975, 66).

There is no empirical evidence, of course, that supports this thesis on the relationship between English, ethnicity and religion in the African context. Yet, many (Afro-Islamic) Hausa and Swahili people that I have had occasion to talk to, both here in the United States and in Africa, are adamant that, everything else being equal, they are "better" speakers of English than members of other ethnic groups in their respective nation-states of Nigeria and Kenya. And so competitive religion and competitive ethnicity in Africa have sometimes met at the political stadium of ex-colonial languages.

ENGLISH AS AN EXIT VISA

In addition to the presumed ethno-religious face of English in Africa, there has been the interplay among English, ethnicity, and gender. The latter partly relates to the language as a possible instrument of temporary "escape" from ethnic-based cultural constraints on the lives of certain categories of members of society. Some African feminists, for example, regard European languages in Africa as both a blessing and a curse, as instruments of liberation on the one plane and vehicles of domination on another. Assia Djebar is of the belief that her entire society stands to lose, often to the advantage of the West, by its "capitulation" to a foreign tongue, in this case the French language. Yet, she continues to believe that the French language provides her with a unique space for self-unveiling, to do with the language what her Arab patriarchal society of Algeria considered taboo for women to do with the Arabic language (Lionnet 1996, 331–333). I have heard similar sentiments expressed on different occasions by women writers from Anglophone Africa in their ambiguous relationship with the English language.

To the extent that it is associated with education, English is also seen to have a liberative potential in a more systemic sense. Throughout Anglophone Africa there are reports of "falling standards" of English, judged of course on the basis of a putative linguistic norm, as the language itself undergoes change in the mill of African social experience. At the same time, however, it has been reported that female students, whose access to the language has generally been more restricted, are increasingly performing better in English-language examinations than their male counterparts. This is certainly true for Kenya (personal communication with Kimani Njogu of the Kenya Examination Council, October 4, 2002).

There are probably several possible reasons for this gender difference in English proficiency among school children. But, in a pilot survey of the subject in the city of Mombasa, Kenya, during the summer of 2000, close to half of the 48 female respondents provided more or less the same explanation for the greater success of their sex in English school examinations: That women were more highly motivated to learn the language because it accorded them new opportunities to escape from their ethnically ascribed status on grounds of their gender.

These results concur with the findings of a South African study on gender and patterns of English usage among Zulu-speaking people, contrasting rural with urban contexts. In the more "traditional" rural setting, where women are regarded as the custodians of ethnic culture, female students are not encouraged to develop too high a proficiency in English. We are told by Dhalialutchmee Appalraju that:

> For a male, it is important to be proficient in English, in that this will give him increased status and furthermore improve his chances in the job market. His proficiency in English is one sign of his success as a male in the community. Females must therefore guard against being too proficient in English, lest they be seen to encroach on male identities. Zulu remains central to female identities, in that women are required to transmit cultural values to children. Retention of Zulu is more important for their identity than developing skills in English. (Quoted by De Kadt 2002, 88)

And in conclusion the study suggests that, through the observance of the restrictions imposed on their acquisition and use of English, females in fact acquiesced to their subordinate status within the ethnic community (De Kadt 2002, 93).

For student respondents in urban schools, on the other hand, the study found that not only do female pupils claim to use English in many more social contexts than male pupils, but that they are also even more convinced than their male colleagues "that English is a far more desirable and important language than Zulu" (De Kadt 2002, 89). From both the South African and Mombasa studies, therefore, we may be witnessing a situation in which, through English, African women are seeking to relocate themselves culturally, challenging the ethnically defined patriarchal boundaries of their identities in new ways.

As in the case of women, gay people in Anglophone Africa may also have found English a useful facilitative tool in their quest to live a gay identity. We know, of course, that, with the exception of South Africa where gays and lesbians are constitutionally protected, male homosexuality is a criminal offence in virtually every African country. In some cases the anti-gay laws have been given the added force of presidential decrees. Many people still remember the verbal onslaught of President Robert Mugabe of Zimbabwe, angered by the sight of a booth of gay and lesbian literature during the 1995 Zimbabwe International Book Fair. As Mugabe declared, "What we [Africans] are being persuaded to accept is sub-animal behavior and we will never allow it here. If you see people parading themselves as lesbians or gays, arrest them and hand them over to the police" (quoted by Dunton and Palmberg 1996, 12–13). Similar homophobic remarks have been made by some other African presidents, including President Daniel Arap Moi of Kenya and President Yoweri Museveni of Uganda.

These laws and presidential sentiments notwithstanding, there are societies in Africa where homosexuality has existed for centuries and, though frowned upon and considered immoral, is definitely tolerated and practiced relatively openly. What is important for our purposes here, however, is that gay people from these communities have not had to rely on English in the performance of their gay identities. They are not constrained to live a gay life only because their linguistic repertoire may be restricted to their own ethnic languages.

There are many Afro-ethnic societies, however, especially in their more rural articulations, where homosexuality is considered a cultural taboo of enormous proportions. Many rural-based gays from these communities, therefore, have to migrate to urban areas to escape the cultural sanctions against their preferred way of life. In addition, it is in the multiethnic urban spaces that they hope to connect with similarly oriented people, usually from other ethnic groups, often from other countries altogether, to belong to a community. Under these circumstances, English is likely to have become an indispensable aid in this attempted escape from the anti-gay ethnic traditions that are particularly prevalent in the rural areas, in search of new spaces to live a gay life. Recent biographical studies of African gay life—e.g., Murray and Roscoe 1998—do suggest a critical role for the English language in the interplay between ethnicity and homosexuality in urban Anglophone Africa.

ENGLISH AS AN ENTRY VISA

If English provides an exit visa, an avenue of escape from certain cultural constraints of one's ethnic group, however, could it be an entry visa into new identities? Earlier on I indicated that the English language exists within certain boundaries that prevent black people from acquiring an Anglo identity even when they speak it as a "native" language. But can English facilitate the integration of black people from one region into black identities in other regions of global Africa? This brings us to the story of new African immigrants to the United States.

The economic havoc wreaked on the African continent by international capital—first in its colonial form and now in its more globalist form, with globalism defined as "the latest stage of imperialism" (Sivanandan 1999)—has led to a continuing outflow of the population, both skilled and unskilled, from Africa to other parts of the world. A 1993 United Nations report indicates that "the world's population now includes 100 million immigrants, of whom only 37 percent are refugees from persecution, war or catastrophe. Migration, that is, is more of an *economic* than a political phenomenon" (quoted by Readings 1996, 48). By all indications, the proportion of refugees from Africa, as the continent worst hit economically, is increasing in leaps and bounds.

What we may be witnessing, then, is a kind of paradox: the economic and cultural Westernization of Africa may be leading to the demographic Africanization of the West, America included. The Westernization of Africa has contributed to the "brain drain" that has lured African professionals and experts from their homes in African countries to jobs and educational institutions in North America and the European Union. The old formal empires of the West have unleashed demographic counter-penetration. Some of the most qualified Africans have been attracted to professional positions in North America and Europe.

But by no means are all African migrants to the West highly qualified. The legacy of colonialism and neocolonialism has also facilitated the migration of less-qualified Africans. Africans, in other words, are growing in numbers at both the top and bottom ends of the vertical pole of social class in the western hemisphere. As expected, many of those entering English-speaking countries such as the United States come from the traditionally Anglophone countries such as Nigeria and Uganda. But there is also an increasing number of migrants from Francophone and Lusophone Africa. Equally significant is the fact that these African immigrants tend to settle in states, cities, and neighborhoods that already have a high proportion of people of African descent and where the racial climate is considered relatively favorable (Takougang 1995, 50–57).

The globalization of Africa is on the rise, as the new African Diaspora, the Diaspora of imperialism—of the dispersed of Africa resulting from the colonial and postcolonial dispensations—grows in numbers. There are already book-length studies of the phenomenon with such telling titles as the "Africanization of New York City" (Stoller 2002). In both absolute and proportional terms, there are more *American Africans* today than at any other point in history. We define American Africans as those immigrants from the continent who have acquired citizenship or residency status in the United States. Partly because of their continued linguistic, cultural, and ethnic linkages with Africa, these members of the *first or migrant generation* of the Diaspora of imperialism tend to be less race conscious than members of the Diaspora of enslavement. This is a difference of orientation that has had its toll on pan-Africanism in the past.

With regard to *American Africans,* in particular, it has been suggested that their conversion to an *African American* identity takes place at precisely the point when they lose their ancestral languages and acquire the English language instead (Mazrui 1999). More significant about this particular section of the Diaspora of imperialism is its potential bridging role. In as much as its members have become "nativized" in their new home in the United States, they continue to have familial connections with the continent of Africa. As a result, they belong to both worlds, so to speak, and are in a position to identify with the immediate concerns, problems, and struggles of both Diaspora.

But through which English-language variety are the various African Diasporas in the United States likely to connect with each other? Will it be through mainstream American English or through Ebonics? The answer may vary from place to place and may be partly dependent on class considerations. Like their middle- and upper-class African American counterparts, the offspring of the "professional class" African immigrants may be more inclined toward some approximation of Standard American English, which may, in turn, foster their mainstream Americanization in a national-cultural sense—even though this national pull within American society may continue to be in competition with the more global pull of pan-African allegiance.

The working-class section of the African Diaspora of neocolonialism, which often ends up sharing the black neighborhood spaces with poorer African Americans, may discover its identitarian links with the Diaspora of enslavement by adding Ebonics to its linguistic repertoire. Through the Ebonics current that regards the variety as an exclusively American-grown medium peculiar to African Americans, diasporized Africans may increasingly experience the pull of black separatist identity within the United States. This linguistic response may be reinforced by the recurrent waves of Anglo-Saxonism in American society and of the offensive against multiculturalism.

In addition to class differences in linguistic paths of African Americanization, there may be a gender gap in the rate of African Americanization of recent African immigrants. Informal discussions with members of the Somali community of Columbus, Ohio, for example, suggest that many parents consider it more important for their female children to retain the Somali language than for their male children. There are also indications that among the first American-born (young) Somali generation, more girls than boys are concerned about the maintenance of the Somali language even as they value the power of the English language in their "new" surroundings. Indeed, the entire project of ensuring that the linguistic umbilical chord with Somalia remains intact seems to have been entrusted to the women more than to the men of this immigrant community. Obviously, this an area that needs further investigation. And, of course, the dialect and rate of African Americanization of African immigrant communities may also be conditioned by other variables such as religion, ethnicity, and national background.

CONCLUSION

In recapitulation, then, among the things that I have tried to demonstrate in this essay is how black perceptions of the linguistic politics of the white "other" have sometimes led black people to make claims about and celebrate the uniqueness of their own varieties of English. This is one face of the two-dimensionality of the black person. And as shown in the case of Ebonics and the new immigrants from Africa, some of these black varieties have become important markers of ethnicity and ethnic shifts within the wider black community. But, in spite of their nationalist selves, Black people often have had to submit to what are seen as "white" varieties of the language to beckon and reach out to each other across boundaries of ethnicity and nation. Sometimes there have been competing claims in the black world, especially in national spaces within continental Africa, about which ethnic group has a better command of the English language than other ethnic groups. This is part of the intra-black dimension of "double-consciousness." But, on closer scrutiny, this dimension is itself conditioned by the black–white dimension to the extent that the degree of linguistic Anglicization, or at least of proficiency in English, has become accepted as a legitimate measure of cultural sophistication.

There is also the interplay among English, ethnicity, and class, again especially as it relates to the African condition. In every Anglophone African country, the English language has been an instrument of communication between different ethnic groups at the upper horizontal level and a linguistic barrier between the elite and the "masses." English has helped erode ethnic behavior (though not necessarily ethnic consciousness) and has accentuated class divisions. It has been at once a force in class formation and a means of "detribalization" in a cultural sense. And, of course, some have found this state of affairs lamentable. As two South African singers have described the new African elite:

> Bits of songs and broken drums
> Are all he could recall
> So he spoke to me
> In a bastard language
> Carried on the silence of guns. (Quoted by Pennycook 1994, 2)

The allusion here is that English has continued to be part of an imperialist arsenal against Africans, at least in the cultural domain.

If English is a tool of detribalization, however, could it also serve as an instrument of black liberation? This, of course, is a subject that has been at the heart of a continuing debate in the black world. Is Audre Lorde correct that the master's tools (English) cannot destroy the master's house (of privilege)? Scholars such as Ngugi wa Thiong'o (1986) would certainly agree with Audre Lorde's proposition. Others will probably lean on the side of James Baldwin that "an immense experience has forged this language; it has been (and remains) one of the tools of people's survival, and it reveals expectations which no white American could easily entertain" (Baldwin 1964, 14).

But as we have seen in the case of gender minorities within ethnic groups, the question of language and liberation has a significance that goes beyond the theoretical. It is, for many of them, a lived struggle of negotiating between the linguistic fetters of ethnic particularism rooted in patriarchy, and those of trans-ethnic universalism, often trapped in Western and elitist terms of reference. Patricia Hill Collins sees this linguistic struggle of black women as one based on rearticulation. In her words:

> . . . rearticulation does not mean reconciling Afrocentric feminist ethics and values with opposing Eurocentric masculine ones. Instead . . . rearticulation confronts them in the tradition of "naming as power" by revealing them very carefully. Naming daily life by putting language to everyday experience infuses it with the new meaning of an Afrocentric feminist consciousness and becomes a way of transcending the limitations of [ethnicity] race, gender and class subordination. (Collins 1991, 111)

There is a sense, then, in which the destiny of black varieties of English that seek a better balance between the imperative of ethnic identities and the quest for black liberation may ultimately be in the hands of the black woman. And it is in that direction that we must begin to focus our attention in the study of language use as we seek to develop a better understanding of the interplay between English and ethnicity in the global African experience.

REFERENCES

Achebe, Chinua. 1965. English and the African writer. *Transition* (Kampala) 4, 18: 26–30.

Ani, Marimba. 1994. *Yurugu: An Africa-Centered Critique of European Cultural Thought and Behavior.* Trenton: Africa World Press.

Asante, Molefi Kete. 1987. *The Afrocentric Idea.* Philadelphia: Temple University Press.

Asante, Molefi Kete. 1989. *Afrocentricity.* Trenton: Africa World Press.

Baldwin, James. 1964. "Why I stopped hating Shakespeare." *Insight* 11 (Ibadan, Nigeria): 11–16.

Blyden, Edward W. 1888. *Christianity, Islam and the Negro Race.* London: W.B. Whittingham.

Chege, Michael. 1997. Africans of European descent. *Transition* 73: 74–87.

Chinweizu. 1988. *Voices from the Twentieth-Century Africa.* Boston: Faber and Faber.

Collins, Patricia Hill. 1991. *Black Feminist Thought: Knowledge, Consciousness and the Politics of Empowerment.* New York: Routledge.

Crummell, Alexander. 1969. *The Future of Africa.* New York: Negro Universities Press.

De Kadt, Elizabeth. 2002. Gender and usage patterns of English in South African urban and rural contexts. *World Englishes* 21, 1: 83–97

Dirven, Rene and Marjolijn Vespoor. 1998. *Cognitive Exploration of Language and Linguistics.* Amsterdam: John Benjamins.

DuBois, W.E.B. 1997. *The Souls of Black Folk*. Boston/New York: Bedford/ St. Martins.

Dunton, Chris and Mai Palmberg. 1996. *Human Rights and Homosexuality in Southern Africa*. Uppsala: Nordiska Afrikainstituet.

Fanon, Frantz. 1967. *Black Skin, White Masks*. New York: Grove Press.

Hawkins, Bruce W. The social dimension of cognitive grammar. In *Discourses and Perspective in Cognitive Linguistics*, W.A. Liebert, G. Redeker, and L. Waugh (eds.). Amsterdam: John Benjamins. 21–36.

Irele, Abiola. 1981. *The African Experience in Literature and Ideology*. Bloomington: Indiana University Press.

Kagame, Alexis. 1956. *La Philosophie Bantu-Rwandaise de L'être*. Bruxelles: Academie Royale des Sciences Coloniales.

Karenga, Maulana. 1978. *Essays in Struggle*. San Diego: Kawaida Publications.

Lionnet, Françoise. 1996. Logiques Métisses: Cultural appropriation and postcolonial representations." In *Postcolonial Subjects: Francophone Women Writers*, Mary Jean Green, Karen Gould, Micheline Rice-Maximin, Keith L. Walker, and Jack A. Yeager (eds.). Minneapolis: University of Minnesota Press, 1996. 321–344.

Lynch, Hollis R. (ed.). 1971. *Black Spokesman: Selected Published Writings of Edward Wilmot Blyden*. London: Frank Cass.

Masolo, D.A. 1994. *African Philosophy in Search of Identity*. Bloomington: Indiana University Press.

Mazrui, Ali A. 1975. *The Political Sociology of the English Language: An African Perspective*. The Hague: Mouton.

Mazrui, Ali A. 1999. Africans and African Americans in changing world trends: Globalizing the black experience. The Inaugural Lecture of the Africa Program, University of Texas at Arlington, February 25.

Mazrui, Ali A. and Alamin M. Mazrui. 1998. *The Power of Babel: Language and Governance in the African Experience*. Chicago: University of Chicago Press.

Murray, Stephen O. and Will Roscoe. 1998. *Boy-Wives and Female-Husbands: Studies of African Homosexualities*. New York: St. Martin's Press.

Ngugi wa Thiong'o. 1986. *Decolonizing the Mind: The Politics of Language in African Literature*. London: Heinemann.

Olaniyan, Tejumola. 1995. *Scars of Conquest/Masks of Resistance: The Invention of Cultural Identities in African, African-American and Caribbean Drama*. New York: Oxford University Press.

Padmore, George. 1956. *Pan-Africanism or Communism?: The Coming Struggle of Africa*. London: Dennis Dobson.

Pennycook, Alastair. 1994. *The Cultural Politics of English as an International Language*. London: Longman.

Pierce, B.N. The author responds. *TESOL Quarterly* 24, 1: 105–112.

Readings, Bill. 1996. *The University in Ruins*. Cambridge: Harvard University Press.

Richards, Dona. 1990. "The implications of African-American spirituality." In *African Culture: The Rhythms of Unity*, M.K. Asante and K. Welsh-Asante (eds.). Trenton: Africa World Press. 207–231.

Saro-Wiwa, Ken. 1992. The language of African literature: A writer's testimony. *Research in African Literatures* 23, 1: 153–157.

Schwarz, F.A.O.1965. *Nigeria: The Tribes, the Nation and the Race.* Cambridge, MA: MIT Press.

Senghor, Leopold Sedar. 1975. The essence of language: English and French. *Culture* 2, 2: 75–98.

Sivanandan, A. 1999. Globalism and the Left. *Race and Class* 40, 1 and 2.

Stoller, Paul. 2002. *Money Has No Smell: The Africanization of New York City.* Chicago: The University of Chicago Press.

Takougang, Joseph. 1995. Recent African immigrants to the United States: A historical perspective. *The Western Journal of Black Studies* 19, 1: 50–57.

Welsh-Asante, Kariamu. Introduction. In *The African Aesthetic: Keeper of the Tradition*, K. Welsh-Asante (ed.). Westport: Greenwood Press, 1993. 1–5.

3

BASIC ENGLISH, CHINGLISH, AND TRANSLOCAL DIALECT

Yunte Huang

"The radio listener," says Walter Benjamin, "welcomes the human voice into his house like a visitor."[1] The human voice that I used to welcome into my house at night when I was a teenager was certainly no ordinary visitor to a small town in southern China. It was the Voice of America. I was eleven, and like most Chinese kids, I had just started learning English in school. One night, I was fiddling with an old, small-size transistor radio that had belonged to my sister. I pulled up the rusted, crooked antenna and switched to the short-wave channels. Turning the knob up and down to search for a channel with bearable audibility—most channels simply buzzed either because my machine was too old or because the signals had been scrambled by the government—I suddenly came to a spot where, after a few seconds of static, a clear, slow, and manly voice in English rang out: "This is VOA, the Voice of America, broadcasting in Special English . . ."

Not surprisingly, this encounter became a crucial point in my bildungromance in the English language. In my ensuing high school years, I regularly tuned in to the daily half-hour broadcast, which began with ten minutes of the latest news followed by twenty minutes of feature programming in American culture, history, science, or short stories. My favorite was the short program called "Words and Their Stories," which introduced American idioms and their colorful etymologies. The broadcasting is called Special English because its vocabulary is limited to 1,500 words, written in short and simple sentences that supposedly contain only one idea, and spoken at a slower pace, about two-thirds the speed of Standard English. Completely oblivious to the ideological agendas propagated by VOA (and also, as I now realize, at the risk of sending my parents to jail, because listening to "politically subversive" foreign radios was illegal at the time and parents would be held responsible for any political "crimes" committed by their pre-adult children), I learned a great deal of English from the broadcasting.

Only years later, when I became a student of literature and started to look closely into the work of some twentieth-century writers, did I begin to see the connections between the VOA programs I had been listening to as a kid and the modern literature I was studying as my field of expertise. VOA's Special English, I learned, was modeled after Basic English, the brainchild of C.K. Ogden and I.A. Richards, coauthors of one of the most important books in modern criticism, *The Meaning of Meaning: A Study of the Influence of Language upon Thought and of the Science of Symbolism* (1925). Ogden was also responsible for the first English translation of Ludwig Wittgenstein's *Tractatus Logico-Philosophicus*, and Richards was arguably the "father" of Anglo-American academic literary criticism. Furthermore, since its inception in 1929, Basic English had drawn the attention of a number of modernists, including Ezra Pound, James Joyce, Louis Zukofsky, Laura Riding, and Wittgenstein. Pound, Joyce, and Zukofsky were all simultaneously fascinated and troubled by the implications of Basic for their modernist poetry and poetics. Riding launched a sustained attack on Basic and its underlying linguistic principles in her magnum opus *Rational Meaning*. And Wittgenstein constantly belittled Ogden and Richards in his lectures and notes.

When my gaze turned to Chinese modernism, however, I was surprised to find that my encounter with Special English and, by implication, with Basic English, was by no means unique to China. History, as opposed to a linear procession, is often a strange palimpsest. Half a century before my encounter with VOA, China had already heard the buzz of Basic English. And like its Anglo-American counterpart, Chinese modernism had also had a strange love–hate relationship with this one-time Esperanto. Moreover, some Chinese writers, such as Lin Yutang, would later immigrate to the United States and become part of the Asian American literary tradition. Their transpacific trajectories further complicate my study of Basic English by making it impossible for me to draw a distinct line between the two bodies of literature and tell stories from so-called both sides—the Anglo-American side and the Chinese side. As opposed to Basic English's desire for debabelization, Lin Yutang's Chinglish and other versions of Asian American pidgin English make a strong case for what I shall call the translocal dialect.

* * *

The conclusion that I came to then was that it seemed impossible to be on both sides of the looking-glass at once. That is, it made me think

how much more dependent one was than one had suspected, upon a
particular tradition of thought from Thales down, so that I came to
wonder how much *understanding* anything (a term, a system etc.)
meant *merely being used* to it . . . . And it seemed to me that all I was
trying to do and that any of the pundits had succeeded in doing, was to
attempt to translate one terminology with a long tradition into another;
and that however cleverly one did it, one would never produce anything better than an ingenious deformation.

T.S. Eliot, letter to I.A. Richards, August 9, 1930

Eliot in his letter was using his own studies in Indian philosophy and
Sanskrit as evidence to cast doubt on Richards's efforts to translate
Mencius and promote Basic English in China. In response to
Richards's invitation to him to visit China and experience Confucian
culture in person, Eliot wrote, "I do not care to visit any land which
has no native cheese." Cultural traditions, then, just like cheese,
would have to be native products before any authentic understanding
could take place; attempts at translation would be equivalent to
desires for occupying an impossible position—"on both sides of the
looking-glass at once," which would produce only an "ingenious
deformation." Sharing Eliot's appreciation for the difficulty of
translation, Richards, however, believed that a solution exists: Basic
English is a tool to combat the "ingenious deformation"; as a universal language, it is a transparent looking-glass that renders both sides
completely visible and communicable to each other.

Basic English was invented by C.K. Ogden in 1929 as an attempt to
"give to everyone a second, or international, language which will take
as little of the learner's time as possible." The word "BASIC" is an
acronym for British, American, Scientific, International, and
Commercial. With a carefully selected vocabulary of 850 words, it is
designed to cover all the essential requirements of communication in
English. Of these 850, the first 100 consist of "operators," including 18
verbs (*come, get, give, go, keep, let, make, put, seem, take, be, do, have, say,
see, send, may*, and *will*) and words such as *if, because, so, as, just, only,
but, to, for, through, yes*, and *no*. There are 400 "general names" such as
copper, cork, copy, cook, cotton; 200 "common things" or "picturables"
such as *cake, camera, card, cart*, and *cat*; and 150 "qualifiers," or
adjectives, such as *common, complex*, and *conscious*.[2]

According to Ogden and Richards, the idea of Basic came from
their collaborated work on *The Meaning of Meaning*. This book was
motivated in part by an idealist desire to prevent the kind of abuse of
language the coauthors had witnessed during World War I. The Great

War was portrayed both in Britain and abroad as largely a war of propaganda, in which the distortion of abstract words such as "freedom," "democracy," and "victory" was a key weapon.[3] They wanted to dispel so-called Word Magic, a relic of a primitive habit of mind by which words substitute themselves for the power of things. Opposing such verbal superstition, Ogden and Richards propose that words are not part of and do not inherently correspond to things, that words "mean" nothing by themselves, and that only when we make use of words do they stand for things and have "meaning." They object to Saussure's notion that meaning is generated by the language system and inseparable from the symbolization process in which a thought (signified) is expressed as a term (signifier). Instead, they cling to a more traditional view of meaning as standing apart from the language in which it is symbolized and insist that a crucial component of meaning does exist in advance of symbolization: the referent, the Thing. In other words, Ogden and Richards see the referent as meaning itself whereas Saussure does not regard meaning as deriving from the referent.[4] Hence they characterize the Saussurean definition of meaning as merely "verbal definition," whereas calling their own "real definition."[5]

The instrumental view of language adopted by *The Meaning of Meaning* would result in the conception of Basic as the application of their theory. As Richards recalled the genesis of Basic, "when [Ogden] wrote a chapter, in *The Meaning of Meaning*, 'On Definition,' at the end of it we suddenly stared at one another and said, 'Do you know this means that with under a thousand words you can say everything?' "[6] In 1929, four years after the publication of *The Meaning of Meaning*, Ogden introduced his first list of Basic vocabulary.[7] He declared that "it is the business of all internationally-minded persons to make Basic English part of the system of education in every country, so that there may be less chance of war, and less learning of languages."[8] Echoing Henry Ford's peace slogan, "make everybody speak English," Ogden suggested that "Basic English for all" was a counterpart of Ford's pacifist prescription for avoidance of another world war: "The so-called national barriers of today are ultimately language barriers. The absence of a common medium of communication is the chief obstacle of international understanding, and therefore the chief underlying cause of war" (*Debabelization*, 13).

Pacifist utopianism aside, this proposal for a language-centered social reform must have appealed to Anglo-American modernists, who, like Ogden and Richards, had also responded to post–World War I cultural fragmentation by rethinking the function of language and imagining the power of poetic language to change the world. But before turning to the modernists, let me delve a bit deeper into the Basic program and explain the process of vocabulary selection that may provide an even stronger link between Basic and Anglo-American modernism, a link manifest in their shared desire for control.

Basic, as Richards put it, "is a technical innovation in the deliberate control of language."[9] Ogden called his method for reducing the size of the vocabulary "Panoptic Conjugation," a term that he had derived from Jeremy Bentham's model prison, the Panopticon.[10] Editor and advocate of Bentham's work, Ogden ascribed to the famous Utilitarianist the inspiration for his own work on Basic. The intellectual debts incurred in two ways: one is the concept of fiction and the other the Panopticon. At the core of Bentham's theory on language lies the notion of fictions, by which he meant the patterns and norms that impute concrete qualities to entities where none exists. The sentence "Music moves the soul" conceals three fictions. Neither "music" nor "soul" is the name of a thing, nor is any physical movement involved in the relation between them, which the sentence is intended to express. Language is forced to introduce fictions by a form of predication, and verbs are especially guilty of composing fictions because of their work in predication, making us talk about qualities as if they were there whereas in fact they are merely linguistic ghosts and bogeys.[11] Compared by Bentham to the serpents of Eden because of their evanescent, slippery meanings, verbs find their population drastically reduced to only eighteen on Ogden's list, and in fact they are no longer called "verbs" but "operators."

Whereas Bentham's concept of fiction equipped Ogden with a theoretical basis for Basic English, the idea of the Panopticon gave Ogden the technique for building the vocabulary list. The Panopticon was Bentham's design for a model prison, a circular building in which the inspector occupies the center and the cells the circumference. By blinds and other devices, the inspector conceals himself from the observation of the prisoners, creating the sense of an invisible omnipresence. The essence of such architectural design is to enable the supervisor to command a perfect, Panoptic view of all the cells, "Panoptic" meaning "all-seeing at a glance."

Ogden developed the Basic vocabulary according to the Panoptic principle. For example, he would put the word "house" in the center of a circle with spokes at the ends of which were *hut, cottage, mansion, bungalow, skyscraper, log-cabin, habitation, residence, domicile, dwelling,* and so on (see figure 3.1). If the center word could, with appropriate adjectives on the Basic list, replace the other words, then the other words were dropped. In this way, the center word occupies what in

Bentham's Panopticon is called the Inspector's Lodge and oversees the other words that are now excluded from "normative" use, or in a sense imprisoned. As Ogden puts it, the Panopticon "enables the entire vocabulary imprisoned in [its] procrustean structure to be envisaged at a glance."[12]

In *Discipline and Publish: The Birth of Prison*, Michel Foucault identifies Bentham's design as a prime example of a modern society that is built upon the principle of discipline and Panopticism. " 'Discipline,' " writes Foucault, "may be identified neither with an institution nor with an apparatus; it is a type of power, a modality for its exercise, comprising a whole set of instruments . . . It is a 'physics' or 'anatomy' of power, a technology . . . We can speak of the formation of a disciplinary society in this movement that stretches from the enclosed discipline, a sort of social 'quarantine,' to an indefinitely generalizable mechanism of 'panopticism.' "[13] The selection of Basic vocabulary enacts exactly such disciplinarism and Panopticism. A brief glance at Bentham's drawing for the Panopticon and an illustration of Ogden's method of vocabulary selection yields a striking visual resemblance as well as the similarity of the mechanisms of control at work in both enterprises.

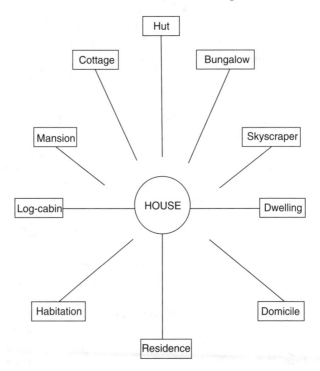

Figure 3.1   An illustration of Ogden's method of vocabulary selection.

Likening the Inspector's Lodge or the center word to the cells or the periphery words is a Panoptic vision or the spoke, which Foucault characterizes as "an uninterrupted work of writing" and Richards sees as the function of "vertical translation."[14] "At the heart of language control," write Richards and Gibson, "is the use of words (or, better, senses) as instruments in looking closely at or into the senses of other words" (*Language Control*, 110). And Richards extends the Panoptic, inspector/prisoner metaphor to describe the relation between words: "This selection, this language within language, can thus serve as a sort of caretaker, an inspectorate, a maintenance, repair and remedial staff, able to examine, criticize, deputize and demonstrate where needed: in brief be a control upon the rest. And not a control merely over its lexical performance, the efficiency of its vocabulary in use, the choice, justice and comprehensibility of its terms. The possible control covers the implications, the requirements and exclusions."[15] As we will see later, this concept of language control will become a complement to Richards's account of poetry as a technical control of meaning and thus provide a linguistic and philosophical justification for the method of "close reading," the hallmark of the New Criticism.

The Panoptic technology is used not only in the building of vocabulary, but also in the teaching of language. On April 30, 1961, the *New York Times* published a story about a mobile classroom designed by Richards and his assistant to aid foreign-language teaching. The so-called Arlington Instruction Van is described by Richards in this way:

> The van itself, the type of trailer employed by a construction company as a building-site office, has positions for a total of 18 students along either wall of the 8-foot wide classroom on wheels. Each student has a clear view of the screen at the front of the van. The instructor, from the rear of the van, controls the film projector and the Inter-Com console through which he can instruct the students individually or as a group. The students hear the tapes and the voice of the teacher through headsets and all is quiet in the acoustically treated instruction van . . . . Since the student's work can be monitored without his realizing it, the instructor is fully able to analyze and correct the student's efforts.[16]

Interestingly, according to the reports by Richards and Gibson on the results of their instructional experiment, the Instruction Van sessions were aimed exclusively at underachieving students, or whom the reports call "problem" boys and girls. The analogy is all-too-obvious between the van segregating academically delinquent pupils and the prison quarantining behaviorally delinquent members of a society.

It is not my purpose here to demonize the use of modern technology for more efficient language teaching. If one looks at some of the pictures of the Arlington Instruction Van, the interior of the mobile classroom is not really that different from that of an ordinary language lab, with cubicle privacy for each student as well as monitoring power for the teacher. But we would be naïve not to see classroom settings as reflections of the mode of production of a society and of the cultural ideology implied in the mode. Jesus, let me remind you, preached from the mountaintop and by the lake; Buddha taught in a garden full of flowers; and Confucius often gave lessons to his disciples at crowded, noisy marketplaces. The location and setup of the instructional venue are inevitably bound up with the conception of knowledge: in the cases of Jesus, Buddha, and Confucius, knowledge is embodied, inseparable from personality, whereas in modern conception, knowledge is disembodied, objective, and instrumental. The degree to which knowledge is disembodied in our modern age can be seen in the very description of the Arlington Instruction Van. As Richards and Gibson explain, one great advantage of using the Van is that "the instruction process no longer hinges solely on a teacher's [linguistic] competence," because the projector and tapes will do the instruction for the teacher. (When I was a student at Peking University, my English Listening and Comprehension class was taught at a language lab and by an instructor whose level of English was not much higher, or maybe even lower, than that of my average classmate.)

As Foucault has reminded us, discipline should not be identified merely with an institution or apparatus; it manifests itself above all as a technology. Bentham's Panopticon, after all, originated from his brother's architectural drawings for a rotunda-shaped workshop in which the laborers are put under complete supervision by an invisible inspector.[17] Hence prisons, workshops, classrooms, and vocabulary lists have all become institutions where the cultural logic of Panopticism is manifested and the technology of control applied.

* * *

your heart would have responded
Gaily, when invited, beating obedient
To controlling hands
. . .
These fragments I have shored against my ruins
. . .
Datta. Dayadhvam. Damyata.

T.S. Eliot, *The Waste Land*

In these lines, which end Eliot's famous poetic response to post–World War I cultural fragmentation, two words stand out: "controlling" and Damyata, Sanskrit for "control" (the Sanskrit triad translated, "Give, sympathize, control"). Eliot, who earlier objected to Richards's efforts in translating Chinese texts and promoting Basic in China, has now come to share with Richards a desire for the control of meaning. Although Eliot uses many foreign phrases and sentences in the poem, including the very Sanskrit word for "control," the appearance of openness, fragmentation, or multilingualism is immediately undercut not only by the thematic coherence of the poem, but also by the use of endnotes by the poet, who apparently wants to aid and ensure proper understanding of the poem. The endnotes thus work as a control mechanism, although the choice of poetic vocabulary veers in the opposite direction from Basic.

It is actually no surprise that despite his objection to Basic, Eliot is Richards's kindred spirit in literary ideology. New Criticism, of which both of them were key founders, is to a large extent predicated on the reader's ability to control textual meaning. New Criticism's notorious distaste for biographical information and historical background, focusing instead on the text itself, had an early rehearsal in Richards's *Practical Criticism*, which was published in 1929, the same year when Basic English was invented. The book was primarily based on the results of experiments he had conducted with his students at Cambridge. He issued printed sheets of poems to his students who were asked to comment freely on them; the authorship of the poems was not revealed and with rare exceptions was not recognized. The students' comments, therefore, would focus only on the texts themselves—a trademark of New Criticism.[18] Such a distaste for contextuality finds its parallel in the kind of decontextualization in Basic.

The other feature of New Criticism, "close reading," is an attempt not only to decontextualize, but also to contain the multiplicity and ambiguity of meaning. In this sense, close reading is a Panoptic technique. But the New Critical Panopticism is manifested even more clearly in Richards's account of poetry, an account according to which poetry *is* Basic English and vice versa. A student of Romanticism, Richards sees poetry as, to quote Coleridge's dictum, "the best words in the best order," that is, the "best language." This "best language," otherwise called "poetic diction" in Romanticism, is a prototype for Basic as "a language within language"; and Richards's accounts of poetry and of Basic are often interchangeable. The technique of poetry, writes Richards, lies in "managing the variable connections between words and what they mean: what they might mean, can't mean, and should mean—that—not as a theoretical study only or chiefly, but as a matter of actual control."[19] Likewise, Basic is "a pioneer prototype for many of the inquiries into symbolic similarities and differences," or a "vertical translation from unrestricted into restricted language."[20] Hence, when Richards maintains that "This capacity of a small segment of the language to exercise such a wide and deep supervision over the rest is the ground for believing that an effective heightening in men's ability to understand one another can—given an adequate attempt—be brought about," we can be quite certain that the "small segment of the language" refers to both poetry and Basic English.[21]

But the New Criticism of Richards and Eliot tells only a partial truth about Anglo-American modernism. If Basic English is a program for decreasing difficulty and ambiguity and New Criticism introduces methods for controlling them, not every modernist shared such a desire for control. On the contrary, as Marjorie Perloff and others have argued, the desire for indeterminacy has been equally strong in the twentieth century.[22] "In the poetry of this 'other tradition,' " writes Perloff, "ambiguity and complexity give way to inherent contradiction and undecidability, metaphor and symbol to metonymy and synecdoche, the well-wrought urn to what Ashbery calls 'an open field of narrative possibilities,' and the coherent structure of images to 'mysteries of construction,' nonsense, and free play."[23] At the very least, many modernists were ambiguous between their desire for control and aspiration for indeterminacy, a fact that is evidenced not only in their work but also in their mixed responses to Basic English.

In 1935, Ezra Pound wrote a review of Ogden's *Debabelization*, a book that seemingly argues against the kind of polyvocality characteristic of Pound's *Cantos*. Pound begins the review with an admission of guilt followed immediately by a self-defense:

> If mere extensions of vocabulary, or use of foreign words is a sin, I surely am chief among all sinners living. Yet, to the best of my knowledge, I have never used a Greek word or a Latin one where English would have served. I mean that I have never intentionally used, or wittingly left unexpurgated, any classic or foreign form save where I asserted: this concept, this rhythm is so solid, so embedded in the consciousness of humanity, so durable in its justness that it has lasted 2,000 years, or nearly three thousand. When it has been Italian or French word, it has asserted or I have meant it to assert some meaning not current in English, some shade or gradation.[24]

On one hand, Pound favored the use of Basic as a means of "weeding out bluffs . . . [and] fancy trimmings," "chucking out useless verbiage," and creating an effect resonant with his Imagistic aesthetics: "Direct treatment of the thing," and "to use absolutely no word that does not contribute to the presentation."[25] On the other hand, as he made clear in the above passage, his poetry taps into linguistic resources spanning continents and ages, a poetic desire that counters Basic's intention to limit and fix the tool of meaning, the tool being not just English but a limited, controlled version thereof. As Pound insists in *ABC of Reading*, "The sum of human wisdom is not contained in any one language, and no single language is CAPABLE of expressing all forms and degrees of human comprehension. This is a very unpalatable and bitter doctrine. But I cannot omit it."[26]

In spite of his advocacy of polyvocality, however, Pound is also notorious for his manifest desire for the control of meaning and value. His pro-fascist ideology has often been interpreted as a symptom for such a desire. Pound's goal is to use as many linguistic resources as possible but also to arrive at a unified picture, a moment of absolute luminosity, or, to use his own term, "the great ball of crystal."[27] Pound's Imagism, with its emphasis on visual clarity, echoes Richards's description of Basic's Panopticism: "*Clear* is one of the key words for any controlled language and we may note here that it has a surprising number of variously relevant senses: bright, unclouded, free from blotches; easily and distinctly heard; able to see or be see distinctly; free from doubt, from guilt; innocent; free from burden, from charges, as in 'clear profit.' "[28]

Speaking of profit, Pound apparently sees a connection between Basic's analytic economy and his own Social Credit theory. The latter, as we know, is a proposal for the control of monetary value by an authoritarian government. It calls for the replacement of paper money by certificates issued by the government as payment for work. Pound believes that in this way social evils, such as usury, which obscures the nature of monetary and linguistic values, could be rooted out. In this sense, usury would be equivalent to what Bentham has condemned as "fiction," a concept that has provided the theoretical foundation for Basic English; and Pound's prescription for social reform is similar to the ones provided by Bentham and his followers, namely Ogden and Richards. Hence, Pound makes a connection between his economic theory and Basic in his review of Ogden's book,

My recent condensed recommendation for Social Credit Policy is as follows:

1. Simplification of terminology.
2. Articulation of terminology.

3. AS MUCH PROPAGANDA AS POSSIBLE SHOULD BE
   WRITTEN IN BASIC ENGLISH.
4. Less tolerance toward converging movements.
5. Hammer on root ideas.[29]

By "articulation of terminology," Pound means the ability to "distinguish the root from the branch," a reference to the Panoptic design of
the Basic vocabulary, in which the center word is the root and the
periphery words are merely branches—"hammer on root ideas." In
his letter to Ogden on January 28, 1935, Pound wrote, "I proposed
starting a nice lively heresy, to effek, that gimme 50 more words and
I can make Basic into a real licherary and mule-drivin' language, capable of blowin Freud to hell and gettin' a team from Soap Gulch over
Hogback. You watch ole Ez do a basic Canto."[30]

The proposed Basic Canto never materialized, but a similar project, one of wedding the simplest language to a literary text whose
linguistic complexity resembles *The Cantos*, did work out, and that is
the Basic version of James Joyce's *Finnegan's Wake*. The Basic
English translation of the last four pages of the Anna Livia Plurabelle
chapter of *Finnegan's Wake* first appeared in Ogden's journal *Psyche*
in 1931 and then was republished the next year in the avant-garde
literary journal *transition*. Joyce's book, as we know, mixes words
from sixty or seventy other languages into its "basically English"
vocabulary, and like Pound's *Cantos*, the novel is excessively allusive
in style, referring to everything from the content of the eleventh
Britannica to popular songs, jokes, and gags culled from comic
books.[31] As Marshall McLuhan put it, "Joyce is making a mosaic, an
Achilles shield, as it were, of all the themes and modes of human
speech and communication."[32] To tame such a linguistically diverse
text, then, would be the ultimate victory for Basic. The result, however, is far from being what Ogden has claimed in the introduction
to the piece, that "the simplest and most complex languages of man
are placed side by side" and that Basic succeeds in being an international language "in which everything may be said."[33] Let's examine
some passages:

> *Joyce's original*: Wait till the honeying of the lune, love! Die eve, little
> eve, die! We see that wonder in your eye. (215)
>
> *Ogden's translation*: Do not go till the moon is up love. She's dead,
> little Eve, little Eve she's dead. We see that strange look in your
> eye. (261)

Joyce: Sudds for me and supper for you and the doctor's bill for Joe John. (215)

Ogden: Washing for me, a good meal for you and the chemist's account for Joe John. (261–262)

Joyce: Flittering bats, fieldmice bawk talk. (215)

Ogden: Winged things in flight, field-rats louder than talk. (262)

Joyce: Tell me, tell me, tell me, elm! Night night! Telmetale of stem or stone. (216)

Ogden: Say it, say it, tree! Night night! The story say of stem or stone. (262)

In the aforementioned review of Ogden, Pound asserts, "If a novelist can survive translation into Basic, there is something solid under his language" (411). In the case of the Basic translation of *Finnegan's Wake*, I leave it to the reader to appraise the success or failure of the translation, to decide whether the poetic effects of "the honeying of the lune, love," of "Tell me, tell me, tell me, elm . . . Telmetale of stem or stone," and the ambiguity between the German "die" and English "die" have all survived translation; or whether "that strange look in your eye," "a good meal," "the chemist's account," and "winged things in flight" sound more like word-riddles than actual translations of "the wonder in your eye," "supper," "the doctor's bill," and "flittering bats," respectively. Joyce's book itself, after all, relies on "punns and reedles" (239). And that may explain Joyce's willingness to cooperate with Ogden on this translation project; that is, rather than seeing the polyvocality of his work absorbed into the neutrality of Basic, Joyce regarded the Basic rendition as a new fragment of the linguistic multiplicity his text intends to include. With its catholic appeal, made possible by its inclusion of something for everyone—a German word here, a French phrase there, even some Chinese pidgin sprinkled into the mix, *Finnegan's Wake* seems to have realized the dream that gave birth to Basic: the dream of a universal language. And it has done so by running the opposite course: to be open to all languages, to rebuild Babel.

"Ogden is against 'Babel,' the confusion of many languages," writes Louis Zukofsky in a 1943 essay on Basic. Earlier in his career, Zukofsky had already experimented with a literary project similar to Basic English. Between 1932 and 1934, he worked on a story entitled "Thanks to the Dictionary," with its vocabulary limited to page samplings from two dictionaries.[34] In his insightful study of

Zukofsky's relation to Basic English, Barret Watten sees "Thanks to the Dictionary" as a reflection of what lies in common between Zukofsky's trademark Objectivism and Basic English: concrete visuality and objectivity of meaning. Watten shrewdly maintains that despite Zukofksy's fondness for poetic objectivity, he was also drawn to the competing aspect of modernist poetics: polysemy.[35] Zukofsky, in his own essay, has already identified the shortcoming of Basic in this respect: "But the refreshing differences to be got from different ways of handling facts in the sound and peculiar expressions of different tongues is not to be overlooked, precisely because they have *international* worth."[36] The word stressed by Zukofsky, *international*, was first coined, as Ogden tells us, by Bentham. If Bentham's internationalism, which comes down to us via Basic, relies on the erasure of differences, the kind of internationalism advocated and practiced by Anglo-American modernists, as we see in Eliot, Pound, and Joyce, draws on those very linguistic differences. The poetic language of these modernist texts is often, to use Joyce's words, "a maundarin tongue in a pounderin jowl" (89). I am not quoting Joyce in vain; Mandarin Chinese in a Poundian jowl has continually fascinated these modernists. If Basic targets Word Magic, the poeticness of the Chinese language, according to these modernists, draws precisely upon it. Zukofsky, in the same essay, tells a story about the magic effect of the Chinese written characters as an antidote to Basic's instrumentalization of language:

> It was a cold winter afternoon toward sunset. The Chinese laundryman had brought back the week's wash and left. When the package was opened, none of his patron's handkerchiefs were in it. The patron walked back in the cold to tell the laundryman. Without looking up the Chinese laundryman said merely: "Go home, you find." "Maybe you come, you find," the patron answered. "All light," the laundryman said gaily. He went out into the cold without bothering to put on a coat and this move troubled the patron.

> In any case, in the house of the man who gives him a week's wash the first act of the Chinese was to go over to the mantelpiece, look at the lot of books and ask: "How much?" "It doesn't much matter," he was told. The laundryman was not interested in looking at the man's linen. "You read English?" the man queried. "No, no savvy."

> The man had another kind of book on his desk shelf, one of the pages opened to a few Chinese ideographs—the characters resembling men standing with legs apart. The English under the Chinese writing read: "Knowledge is to know men; Humanity is to love them." the man

thrust the book onto the laundryman, who responded gaily: "Heh, heh, yeh, handkerchiefs tomorrow!"

Zukofsky, draws a moral lesson from the story, with a jab at Basic:

> Evidently the Chinese was not interested in the handkerchiefs that day. And the other man was not a little impressed by the effect on the Chinese of a force that might be sensed as active in the Chinese characters. At any rate, something more active than the man could find that day in a list of 400 general things and 200 picturable. (156–157)

The notion of "a force that might be sensed as active in the Chinese characters" would be conceived by Ogden as verbal superstition. But Anglo-American modernists did believe in such a magic force at work in the Chinese characters, "something more active" than Basic words. Pound was no Chinese laundryman, and neither was Zukofsky. But the former made a career out of his dealings in Chinese and the latter founded a school of Objectivism, which treats words like objects just as the Chinese characters are regarded as natural signs. The question is: Do Chinese themselves actually believe in the alleged magic of their language? And what happened when Chinese came into contact with a language like Basic, which regards Word Magic as its enemy? To answer these questions, we need to turn to Chinese modernism and witness its encounter with Basic English.

* * *

> A better medium should, from the beginning, recognize that disparity (due to differences between Chinese and Western intellectual traditions) between Chinese and Western attitudes to language and its meaning . . . It should aim at giving the Chinese learner of English what his own language does not (and perhaps never will) provide him with, an instrument of analytical discrimination between meanings . . . The only way in which false and misleading approximations to Western units of meaning with Chinese "equivalents" can be avoided is by giving these meanings through, and together with, an apparatus for comparing meanings—through an explicit analytic language. Such a language is *Basic English.*
>
> I.A. Richards, Basic in Teaching: East and West

The introduction of Basic English in China began with Richards's arrival at Tsing Hua University in Peking in 1929. He had been invited as Visiting Professor to teach freshman English and other subjects there. In a paradoxical way, Richards's Basic enterprise could not have been launched at a better or worse time in China. The Chinese language reform movement, which had begun in the late nineteenth century, was entering a new era in the late 1920s and early 1930s. The phoneticization of Chinese called for by native Chinese scholars would have dovetailed with the adoption of an imported alphabetic language like Basic. But the problem lies exactly in that it is an imported product, and the Chinese response to Basic reveals the Janus face of Chinese modernity: its simultaneous aspiration and resistance to the West.

Although the phoneticization of Chinese had started in as early as 1605 when the Jesuit missionary Matteo Ricci wrote a book in which he annotated Chinese texts with pronunciations in the Roman alphabet, it was not until the late nineteenth century that, as a result of increasing contact and conflict with the West, the Chinese were compelled to rethink the nature of their language and its correlation with the future of Chinese civilization in the world. Diametrically opposite to Anglo-American modernists' idealization of the Chinese written characters, Chinese modernists saw the script as responsible in part for the backwardness of their culture. The lack of a correspondence between writing and speaking, they charged, has created an insurmountable obstacle for developments in rationality, science, and technology, developments directly needed in order to resuscitate China. Unlike their later, more sophisticated views, the proposals they had made in the first two decades of the twentieth century were strikingly radical. Qian Xuantong, for instance, a key player in the new cultural movement, called for a total abolishment of the Chinese language and the adoption of an alphabetic world language as a lingua franca in China. Qian explained his rationale in this way:

> To abolish Confucianism and to eliminate Taoism is a fundamental way to prevent the fall of China and to allow the Chinese to become a civilized nation in the twentieth century. But a more fundamental way than this is to abolish the written Chinese language, in which Confucian thought and fallacious Taoist sayings are recorded.[37]

Less radical proposals called for the creation of a system of "symbols for phonetic notation" that would parallel the Chinese script, or of a Latin system of spelling that would replace the traditional script. The rationales behind these proposals remained the same: the Chinese language is outdated and therefore needs to be either modernized or abolished.

Such a view would actually fit well with the rationale for promoting Basic in China. As seen in this section's epigraph from Richards, Chinese is regarded as a linguistic instrument that is, unlike English, incapable of analytically discriminating between meanings. In *Debabelization*, Ogden quotes a Chinese scholar's characterization of the language to support his own cause:

> Dr. Yen points out that the Chinese language itself is very defective from the standpoint of clearness, accuracy, and logical consequence. "It is a language more appropriate for the expression of poetical and literary fancies than for the conveyance of legal and scientific thought." Time, place, and mode have to be largely implied, or left to the reader to supply. All this, of course, is apart from the well-known absence of scientific terms. (132–133)

Richards identifies another problem, that is, the Chinese attitude toward language and meaning. As Richards explains in the essay "Sources of Conflict":

> The root difficulty is that the fundamental Chinese attitude to statements is unlike that attitude to statements which in the West led to the development of an explicit logic and of that critical reflective examination of meanings which had produced modern scholarship. In brief, the difference is this: The modern Western scholar . . . devotes himself, first, to determining (as neutrally, consciously and explicitly as possible) what the meaning of a passage is, and second, to discussing by an open and verifiable technique whether it is true or false. But traditional Chinese scholarship has spent its immense resources of memory and ingenuity upon *fitting the passage into* an already accepted framework of meanings.[38]

Richards considers "this tendency to accommodating interpretations" as a formidable obstacle to understanding, resulting in that "the studies made by Chinese in Western subjects do not in general as yet give them . . . that power of critical neutral examination and understanding which should be their prime purpose." The solution to this problem, Richards believes, is for the Chinese to gain knowledge of Western ideas directly through a Western language rather than through Chinese (mis)translation. Basic English, since it is easy to learn and has a controlled vocabulary, emerges as the best candidate for this purpose.[39]

In the years after his arrival in Peking in 1929 and before the outbreak of the Sino-Japanese War in 1937, Richards, with the help of his Western and Chinese colleagues at Tsing Hua University, promoted his Basic program quite successfully in China. With grants from the Rockefeller Foundation, which was pursuing its own interests in China at the time, Richards was able to establish the Orthological Institute of China, making connections with the similar institutes that Ogden had founded in Britain, India, Japan, and other countries as regional headquarters for Basic English. The Institute published Richards's Basic textbook, *First Book of English for Chinese Learners*. A number of Chinese universities and middle schools adopted Basic into their curricula. In May 1937, Richards met with the Minister of Education and a government-appointed committee and successfully obtained their approval of his program for the teaching of Basic in middle schools nationwide. Were it not for the Japanese invasion two weeks later, which disrupted the work of the Chinese government and led to the abortion of the original plan, the fate of Basic in China might have been a different story.

But attributing the failure of Basic in China to an unforeseen historical event may only be wishful thinking. Before its demise in China, Basic had already run into obstacles created by a large number of Chinese modernists who aspired to the West on one hand but remained loyal to the Chinese language on the other. Among them, Lin Yutang stood out as perhaps the most articulate opponent of Basic. In his November 16, 1933, letter to Richards, R.D. Jameson, director of the Orthological Institute in China, reported on the critical campaign that was being waged by some Chinese writers against Basic. Jameson cited Lin as the "leader of the Anti-Basic Movement" in the Chinese press. He did not provide Richards, who was back in Britain at the time, with details of Lin's objections on the grounds that "from his English articles it does not seem to me that his opposition is particularly serious."[40] But Jameson had apparently been fooled by Lin's idiosyncratically lighthearted, self-mocking style of prose. As I have argued elsewhere, behind Lin's humor lies his most profound critique of the West.[41]

The articles Jameson referred to were Lin's " 'Basic English' " and "In Defense of Pidgin English," published a few months earlier in the Chinese-run English weekly, *The China Critic*. At the beginning of the first article, Lin seems willing to acknowledge some merits of Basic: "there is no question of the essential value of such a wise selection of vocabulary for people who must get along with what they have time for and who do not aspire to go into the niceties of the English language."[42] But he is quick to identify problems with Basic and here he unleashes his sharp barbs of satire that will later earn him a great reputation in the United States with his English bestseller, *My Country and My People* (1935). Lin points out that because of the limitation of vocabulary, writing in Basic will inevitably fall into utter circumlocution: "The most fervent image of imagination" becomes "the most burning picture that has existence only in mind"; a "beard" becomes "growth of hair on the face"; and a woman's "breast" becomes a "milk vessel."[43] In terms of humor, however, nothing beats the restaurant menu that Lin designed by using Basic vocabulary:

A BASIC MENU

False soup of swimming animal with round

hard cover

or

Soup of end of male cow

Fish with suggestion of China or the

Peking language

Young cow inside thing nearest the heart

boiled in oil

Fowl that has red thing under mouth, that makes

funny, hard noise and is eaten by Americans on

certain day, taken with apple cooked with sugar

and water, but cold

meat with salt preparation that keeps long time

Hot drink makes heart jump or you don't go to sleep[44]

Imagine yourself sitting down in a Chinese restaurant in the United States and being presented with such a Basic Menu. This imagined comical situation may only be compared with the one in which we face the crazy Chinese encyclopedia quoted by Foucault in the Preface to *The Order of Things*. In this encyclopedia, "animals are divided into: (a) belonging to the Emperor, (b) embalmed, (c) tame, (d) sucking pigs, (e) sirens, (f) fabulous, (g) stray dogs, (h) included in the present classification, (i) frenzied, (j) innumerable, (k) drawn with a very fine camelhair brush, (l) et cetera, (m) having just broken the water pitcher, (n) that from a long way off look like flies." The Basic Menu mocks the universalist desire to describe culture-specific objects such as turkey and coffee in a different cultural context by adopting a pseudo-universal language. Likewise, Foucault's Chinese encyclopedia draws a cultural relativist lesson out of laughter. "The wonderment of this taxonomy . . . the exotic charm of another system of thought," writes Foucault, "is the limitation of our own, the stark impossibility of thinking *that*." It is no wonder that Basic has drawn criticism from major proponents of linguistic and cultural relativism such as Ludwig Wittgenstein and Benjamin Lee Whorf. The former has famously said that "to imagine a language means to imagine a form of life" and the latter maintains, along with Edward Sapir, that our understanding of the world is conditioned by our own linguistic structure. But Lin's critique of Basic is not based simply on the grounds of linguistic relativism. His advocacy of Pidgin, I argue, has outgrown the theoretical framework of linguistic relativism by projecting, in a manner not too dissimilar to the *Cantos* or *Finnegan's Wake*, a world of cosmopolitan polyvocality, by reinventing a translocal dialect that has no single, identifiable cultural origin.

As an alternative to Basic, Lin advocates Pidgin, which is a mixture of English and Chinese, or what I would call "Chinglish." According to Lin, Chinglish has at least three advantages over Basic. First, it is much more expressive, as Lin seconds Bernard Shaw's opinion that the pidgin "no can" is a more direct and forceful expression than the "unable" of Standard English. "When a lady says she is 'unable' to come, you have a suspicion she may change her mind and perhaps come after all, but when she replies to your request with an abrupt, clear-cut 'no can,' you know you have to reckon without her company." Second, it will have a brighter future than Basic because of its wide base of support: "Advocates of English as an auxiliary international language have often advanced as an argument in its favor the fact that the language is now spoken by over five million people. By this numerical standard, Chinese ought to stand a close second as an international language, since it is spoken by four hundred fifty million, or every fourth human being on earth." Therefore, a mixture of Chinese and English will defeat any language as the lingua franca of the world. Third, if being analytic is a prerequisite for an international language, Chinglish is more analytic than Basic: "The trouble with Basic English is that it is not analytic enough. We find the word 'gramophone,' for instance, circumlocuted in Basic English as 'a polished black disc with a picture of a dog in front of a horn.' In 2400 A.D., we could call it more simply in real pidgin as 'talking box.' " Likewise, "telescope" and "microscope" can simply be called "look-far-glass" and "show-small-glass"; "telegraph," "electric report"; "telephone," "electric talk"; "cinema," "electric picture"; and "radio," "no-wire-electricity."[45]

It is no wonder that, because of Lin's idiosyncratic prose style, Jameson had made light of his objections to Basic. But the notion that the pidginized "look-far-glass" and "no-wire-electricity" are better expressions than "telescope" and "radio" of Standard English was a shared belief among Lin's fellow Chinese writers. T.F. Chu, in his essay "This Easy Chinese Language," published in the August 31, 1933, issue of the *China Critic*, also uses these two and other pidgin expressions as examples of Chinese's superiority over English. These new terms that have come into being during China's contact with the West, Chu writes, were all coined by using the principle of "expediency" of the Chinese language.[46] By "expediency," Chu apparently refers to what the linguist Otto Jespersen has characterized as Chinese's capacity for freely and regularly combining short elements of a phrase or sentence. This capacity, Jespersen argues in his influential *Progress in Language* (1894), places Chinese in an advanced stage of linguistic progression, more advanced even than English. Both Chu and Lin concur with Jespersen's thesis, as Lin writes in "In Defense of Pidgin English,"

> The whole trend of the development of the English language teaches us that it has been steadily advancing toward the Chinese type. English has triumphed over grammatical nonsense and refused to see sex in a tea cup or a writing desk, as modern French and German still do. It has practically abolished gender, and it has very nearly abolished case. It has now reached a stage where Chinese was perhaps ten thousand years ago. (55)

And Lin goes on to say that "James Joyce and pidgin English will do the rest and complete that historical process until English is as simple and as logical as Chinese" (48). Based on such a comparison, Chu even suggests that an equivalent to Basic English be created in Chinese, perhaps in the hopes of making Basic Chinese, rather than Basic English, the international language (856).

Chu's proposal seems to reveal the nationalist sentiments embedded in the campaign against Basic English, but the issue is more complicated. At the time when Basic was being promoted in China, the advocates of Chinese-language reform had already given up their earlier, more radical stances, such as abolishing Chinese altogether and adopting Esperanto or French as China's official language. In the late 1920s and early 1930s, Chinese reformers, including Lin Yutang, had concentrated on two proposals: *guoyu luomazi* and *latinxua sin wenz*. Both proposals took a pragmatic and ambivalent approach to China's linguistic modernization: they call for, on one hand, abandoning the Chinese script, which has been the bastion of traditional Chinese culture; but they want to preserve, on the other hand, the Chinese language in its spoken form, preventing any potential takeover by a foreign language, such as Basic English. Such an insistence on the vocal, the vernacular, even as the written script is being replaced in a wholesale manner by a Western alphabet, not only reveals the excruciating pain accompanying China's social modernization, but also raises an interesting perspective on the alternative roads that lead to different cultures' linguistic modernities.

I call Chinglish a "translocal dialect" because it not only transcends geographical boundaries, but also unsettles the putative connection between a dialect and a localized, romanticized origin. Unlike Fukienese, Lin's native dialect, or Cantonese, the other dominant dialect among Chinese Americans, which often functions as a natural bond for the immigrant community, Chinglish is an invented vernacular in the sense that it only resembles various versions of pidgin that are used in real life. In fact, Chinglish, as Lin imagined, exists only as a literary language, which is not to say that it has no sociological basis or has no effect in real life—as if literature were not part of real life. In literature, the use of a specific language and style is often a result of a conscious decision made by the writer. But I am more interested in the creation of a particular literary code than the adoption of a preexisting one.

In his essay "Poetics of the Americas," Charles Bernstein distinguishes between dialectical writing and ideolectical writing. By dialectical writing, he means a language practice that refuses allegiance to Standard English but still bases its norm on an affiliation with a definable group's speaking practice. By ideolectical writing, he refers to an ideologically informed nonstandard language practice that rejects both Standard English and any localized, group-based linguistic norm. "Dialect," writes Bernstein, "has a centripetal force, regrouping often denigrated and dispirited language practices around a common center; ideolect, in contrast, suggests a centrifugal force, moving away from normative practices without necessarily replacing them with a new center of gravity."[47] Chinglish may be regarded as an example of ideolectical writing, which, as Bernstein insists, has no easily identifiable marker of group identity or authenticity. And that may indeed explain why for many years Lin Yutang has been criticized by the canon-makers of Asian American literature for adopting a seemingly lighthearted, Chinglish style of writing, a style they believe to be symptomatic of his capitulation to the stereotype imposed on Asian Americans, as weak-minded, incompetent speakers of English. What they have missed is not only the critical edge of Chinglish against linguistic standardization, an issue to which I will turn in a minute, but also the significant way in which literature engages social reality, not by means of representation or reinforcement of identitarian representation, but by exploring the possibilities of such representation, refusing to be bound to the restrictions of rationalized ordering systems. Literature can be, in Bernstein's words, "a process of thinking rather than a report of things already settled; an investigation of figuration rather than a picture of something figured out" (117). The practice of dialectical writing may have its tremendous political edge against linguistic standardization, but its centripetal pull toward a new center reminds us of the very trap into which part of Anglo-American modernism is falling. As I discussed earlier, despite its desire for openness and fragmentation, Anglo-American modernism also has a penchant for control. If by "nonstandard" we only mean different but controllable, then I would rather it be different and exploratory.

Having discussed how Chinglish as a translocal dialect deviates from dialectism's local norms, I would now like to address how it deflates Basic English's global dreams. Chinglish, in short, goes against the grain of Basic English in two ways. First, it constitutes a Chinese response to English's linguistic imperialism, a response that originates in part from nationalism. But I am more interested in the second aspect, in Chinglish as a critique of English not from without but from within. That is, the question of Chinglish is not simply an issue of China versus the West, but a Chinese American issue. Basic, as I said, is a "controlled" language, and the word "control" should be understood by its etymology, "to check or verify, and hence regulate; or to check by comparison, and test the accuracy of" (*OED*). In other words, Basic is an extreme version of standardization, resembling in essence a project of linguistic purification that gained great momentum in modern Anglo-America. The publication of the *Oxford English Dictionary* was the best example of linguistic purification and codification against the onslaught on English by immigrants who flooded into the colonial centers and by the colonial subjects who had adopted and, in the eyes of the purists, "abused" the colonial language. Basic shared such a fear of contamination. In *Practical Criticism*, a book that paved the road for the founding of New Criticism, Richards already expressed concerns with the "decline in speech," which he believed was caused by the increased size of "communities" and the mixtures of culture. "We must," writes Richards, "defend ourselves from the chaos that threatens us by stereotyping and standardizing both our utterances and our interpretations. And this threat, it must be insisted, can only grow greater as world communications, through the wireless and otherwise, improve." And he repeated such a line of reasoning in *Basic English and Its Uses*, suggesting that "Basic English, by providing invulnerable but adequate substitutes for [those] more delicate instruments, can serve our language as a fender. It can guard full English from those who will blur all its lines and blunt all its edges if they try to write and talk it before they have learned to read it."

When such a fear of linguistic contamination reaches an extreme, even Basic itself will be regarded as a potential danger to English. One objection that came from Western linguists was that Basic runs the risk of becoming a pidgin. Pidgin, by definition, "represents a language which has been stripped of everything but the bare essentials necessary for communication. There are few, if any, stylistic options. The emphasis is on the referential or communicative rather than the expressive function of language."[48] These features of pidgin eerily resemble those of Basic. Hence, F.R. Leavis, who extolled Richards as a *locus classicus* in literary criticism and a leader in the elitist campaign against popular culture, had this to say about Basic in *Mass Civilization and Minority Culture* (1930):

> No one aware of Shakespeare's language can view quite happily the interest taken by some of the most alert minds of our day in such a scheme as "Basic English." This instrument, embodying the extreme of analytical economy, is, of course, intended for a limited use. But what hope is there that the limits will be kept? If "Basic English" proves as efficacious as it promises it will not remain a mere transition language for the Chinese. What an excellent instrument of education it would make, for instance, in the English-speaking countries! And, if hopes are fulfilled, the demand for literature in "Basic English" will grow to vast dimensions as Asia learns how to use this means of access to the West. It seems incredible that the English language as used in the West should not be affected, especially in America, where it is so often written as if it were not native to the writer, and where the general use of it is so little subject to control by sentimental conservatism.[49]

I want to flatter myself by thinking that the frightening American scene Leavis alerts us to would include me and my writing at this moment in a language that *is*, rather than "as if it were," not native to me. In this sense, Leavis was quite prophetic, because after all, I *am* a product of Basic English, I have used Basic I learned from VOA's Special English programs to access the West, and now I am trying in my however limited way to tinker with Shakespeare's language. But I am humbled by the realization that my tinkering has not been as successful as what was done by those immigrants chastised by Henry

James in his famous 1905 lecture, "The Question of Our Speech." " These immigrants, James said, "play, to their heart's content, with the English language, or in other words, dump their mountain of promiscuous material into the foundations of the American."[50] James's "mountain of promiscuous material" reminds me of what Milton Murayama has characterized as the "shit pyramid" in Hawaii. In *All I Asking for Is My Body*, Murayama describes a pyramidal structure that is at once monetary (different people receive wages according to different pay scales), spatially sanctioned in the layout of the plantation (a tiered housing system in which descending levels of the pyramid housed different ethnic groups), and linguistic (a scale that descends variously from Standard English to pidgin English, from standard Japanese to pidgin Japanese, etc.). Even shit was organized according to the plantation pyramid, hence the term "shit pyramid."[51] But such linguistic stratification, which would have pleased James, is no longer stable. Let me refer you to a poem by the Hawaii-based Japanese American writer Lois-Ann Yamanaka. Yamanaka is well known for her use of pidgin. In this poem, entitled "Tita: Boyfriends," the teenage speaker switches between Hawaiian creole English and what might be called standard American California Valley Girl English (one can hear that accent in Yamanaka's oral performance of the poem):[52]

Boys no call you yet?
Good for you.
Shit, everyone had at least
two boyfriends already.
You neva have even *one* yet? ·
You act dumb, ass why.
All the boys said you just one little kid.
Eh, no need get piss off.

Richard wen' call me around 9:05 last night.
Nah, I talk *real* nice to him.
Tink I talk to him the way I talk to you?
You cannot let boys know your true self.
Here, this how I talk.
Hello, Richard. How are you?
Oh, I'm just fine. How's school?
My classes are just greeaat.
Oh, really. Uh-huh, uh-huh.
Oh, you're so funny.
Yes, me too, I love C and K.
Kalapana? Uh-huh, uh-huh.

He coming down from Kona next week.
He like me meet him up the shopping center . . .

One of the effects of the dramatic code-switching is that the standard and the creole languages become opaque to each other: neither can the former claim to be the "better" language, one that stands at the top of the shit pyramid, nor can the latter celebrate its often-romanticized authenticity of local color. In other words, both become marked, restricted languages, in the same way as Basic English loses its transparency as a lingua franca and runs the risk of becoming merely a pidgin, or another translocal dialect.

* * *

> We have lingered in the chambers of the sea
> By sea-girls wreathed with seaweed red and brown
> Till human voices wake us, and we drown.
> T.S. Eliot, "The Love Song of J. Alfred Prufrock"

> till other voices wake
> us or we drown
> George Oppen, "Till Other
> Voices Wake Us"

Against Eliot's scene of awakening and drowning by human voices and in the spirit of Oppen's resolute revision, I want to describe another scene, not again of my listening to VOA, but after it. On hot summer nights in the south, I often slept outside our house, on a bamboo bed set up by the cobblestone street and covered with a white, translucent mosquito net. Before dawn, I was always awakened by the noise of fruit farmers bargaining with traders at the nearby market. As typical of southern China's linguistic diversity, these people used at least three dialects to communicate with each other, dialects that were not all known to me. At such moments, hovering between the worlds of dream and reality, I was often overcome by a weird feeling: that I had just woken up in a foreign land, where its people spoke in foreign tongues. If VOA has transported me to a world of cosmopolitanism that lived only in my prepubescent imagination, the babbling noise from a local market had already revealed to me the truth about the cosmopolitanism that characterizes the world in which I wish to live. It is a world that speaks, if I may quote Joyce again, "a maundarin tongue in a pounderin jowl."

NOTES

1. Walter Benjamin, "Reflections on Radio," in *Selected Writings*, Vol. 2, 1927–1934, trans. Rodney Livingstone and others, ed. Michaels W. Jennings et al. (Cambridge, Harvard University Press, 1999), 544.
2. C.K. Ogden, *Basic English: International Second Language* (New York: Harcourt, Brace & World, Inc., 1968), 5–95.
3. John E. Joseph, "Basic English and the Debabelization of China," in *Intercultural Encounters—Studies in English Literatures*, ed. Heinz Antor and Kevin L. Cope (Heidelberg, Germany: Universitätsverlag C. Winter, 1999), 52.
4. Joseph, "Basic English," 60.
5. Joseph, "Basic English," 60. John Paul Russo, *I. A. Richards: His Life and Work* (Baltimore: Johns Hopkins University Press, 1989), 110. C.K. Ogden and I.A. Richards, *The Meaning of Meaning: A Study of the Influence of Language upon Thought and of the Science of Symbolism* (New York: Harcourt, Brace & Company, Inc., 1923), 208 (hereafter cited as *Meaning*).
6. I.A. Richards, "Beginnings and Transitions" (interview by Reuben Brower), in *I. A. Richards: Essays in His Honor*, ed. Reuben Brower, Helen Vendler, and John Hollander (New York: Oxford University Press, 1973), 34.
7. *Psyche* 9, 3 (1929).
8. C.K. Ogden, *Debabelization* (London: Kegan Paul, 1931), 12.
9. I.A. Richards, *So Much Nearer: Essays toward a World English* (New York: Harcourt, Brace & World, Inc., 1968), 248.
10. C.K. Ogden, editorial, *Psyche* 9, 1 (1928), 2–3.
11. W. Terrence Gordon, "Undoing Babel: C.K. Ogden's Basic English," in *Etc: A Review of General Semantics* 45, 4(1988), 338.
12. C.K. Ogden, *Basic English and Grammatical Reform*, supplement to *Basic News* (July 1937), 10; quoted in Russo, *I. A. Richards*, 768.
13. Michel Foucault, *Discipline and Punish: The Birth of the Prison*, trans. Alan Sheridan (New York: Vintage Books, 1979), 215–216.
14. Foucault, *Discipline and Punish*, 197. I. A. Richards and Christine Gibson, *Techniques in Language Control* (Rowley, MA: Newbury House Publishers, 1974) 55.
15. Richards, *So Much Nearer*, 242.
16. The 4-page typescript, entitled "Memorandum for the New York Times," is now in the Richards collection at the Harvard University Archive, Box 2.
17. Mack, *A Bentham Reader*, 189.
18. I.A. Richards, *Practical Criticism: A Study of Literary Judgment* (New York: Harcourt, Brace & Company, 1929), 3.
19. Richards, *So Much Nearer*, 175.
20. Richards, *So Much Nearer*, 49.
21. Richards, *So Much Nearer*, 242.

22. Marjorie Perloff, *The Poetics of Indeterminacy: Rimbaud to Cage* (Princeton: Princeton University Press, 1981; reprint, Northwestern University Press, 1999); Peter Quartermain, *Disjunctive Poetics: Gertrude Stein and Louis Zukofsky to Susan Howe* (New York: Cambridge University Press, 1992).

23. Perloff, *Poetics of Indeterminacy*, back cover.

24. Ezra Pound, "Debabelization and Ogden," in *The New English Weekly* (February 28, 1935), 410.

25. Pound, "Debabelization and Ogden," 411. Ezra Pound, *Selected Letters: 1907–1941*, ed. D.D. Paige (London: Faber and Faber, 1950), 267. Ezra Pound, *Literary Essays of Ezra Pound*, ed. T.S. Eliot (London: Faber and Faber, 1954), 3.

26. Ezra Pound, *ABC of Reading* (New York: New Directions, 1951), 34.

27. Ezra Pound, *The Cantos* (New York: New Directions, 1970), 815.

28. Richards, *Language Control*, 6–7.

29. Pound, "Debabelization and Ogden," 410.

30. Pound, *Selected Letters*, 266.

31. John Bishop, Introduction to James Joyce, *Finnegan's Wake* (New York: Penguin Books, 1999), vii–viii; subsequent quotations from *Finnegan's Wake* are all from this edition.

32. Marshal McLuhan, *The Gutenberg Galaxy: The Making of Typographic Man* (Toronto: University of Toronto Press, 1962), 75.

33. C.K. Ogden, "James Joyce's Anna Livia Plurabelle (in Basic English)," *transition* 21 (1932), 259.

34. Quartermain, *Disjunctive Poetics*, 13–14.

35. Barret Watten, "New Meaning and Poetic Vocabulary: From Coleridge to Jackson Mac Low," *Poetics Today* 18, 2 (Summer 1997), 147–186.

36. Louis Zukofsky, "Basic," in *Prepositions +: The Collected Critical Essays*, foreword by Charles Bernstein (Hanover and London: University Press of New England, 2000), 163; emphasis in original.

37. Qian Xuntong, "Zhongguo jinhouzhi wenzi wenti," quoted in Q.S. Tong, "The Bathos of a Universalism: I.A. Richards and His Basic English," in *Tokens of Exchange: The Problem of Translation in Global Circulations*, ed. Lydia H. Liu (Durham: Duke University Press, 1999), 339.

38. Richards, *So Much Nearer*, 223–224; italics in original.

39. Richards, *So Much Nearer*, 224–225.

40. R.D. Jameson to I.A. Richards, November 16, 1933; quoted in Rodney Koeneke, "Empires of the Mind: I.A. Richards and Basic English in China, 1929–1979" (Ph.D. diss., Stanford University, 1996), 145.

41. See my discussion of Lin Yutang's criticism of Basic and defense of Pidgin in my *Transpacific Displacement: Ethnography, Translation, and Intertextual Travel in Twentieth-Century American Literature* (Berkeley: University of California Press, 2002), 126–133.

42. I am quoting from the reprints of the two essays in Lin Yutang, *The Little Critic: Essays, Satires and Sketches on China* (*Second Series: 1933–1935*) (Westport, CN: Hyperion Press, 1983), 50.

43. Yutang, *The Little Critic*, 52–53.
44. Yutang, *The Little Critic*, 59.
45. Yutang, *The Little Critic*, 54–57.
46. T.F. Chu, "This Easy Chinese Language," *The China Critic* 6, 35 (1933), 856.
47. Charles Bernstein, *My Way* (Chicago: University of Chicago Press, 1999), 121.
48. Suzanne Romaine, *Pidgin and Creole Languages* (New York: Longman, 1988), 24.
49. F.R. Leavis, *Mass Civilization and Minority Culture* (Cambridge, England: The Minority Press, 1930), 29–30.
50. Henry James, *The Question of Our Speech* (Boston: Houghton Mifflin, 1905), 42–43.
51. Milton Murayama, *All I Asking for Is My Body* (Honolulu: University of Hawaii Press, 1988), 28.
52. Lois-Ann Yamanaka, *Saturday Night at the Pahala Theatre* (Honolulu: Bamboo Ridge Press, 1993), 35.

PART 2

REPRESENTATIONS

4

REPRESENTING JEWISH IDENTITY THROUGH ENGLISH

Cynthia Goldin Bernstein

The term *Jewish English* (JE) has come to refer to a variety of the English language influenced by Hebrew and Yiddish and spoken primarily by American Jews of Eastern European origin. Although some features of this complex variety have entered into the American mainstream, JE, like other ethnic varieties, serves primarily to represent affiliation with a shared cultural heritage. At the same time, Jewishness does not mean the same thing to all those who identify themselves as Jewish: members differ with respect to religious observances, holiday rituals, national origins, political views, places of residence, and so on. All of these factors influence the way people speak. This essay investigates how ethnic identity has affected the history and development of JE; how JE varies from general American English with respect to vocabulary, grammar, pronunciation, and discourse style; and how JE represents the struggle for the dual identity of being Jewish and being American.

JEWISH ETHNIC IDENTITY AND THE DEVELOPMENT OF JEWISH ENGLISH

Ethnic groups, according to the National Council for the Social Studies, fall into three general categories: those "distinguished primarily on the basis of race, such as African Americans and Japanese Americans"; those "distinguished on the basis of national origin, such as Polish Americans"; and those "distinguished primarily on the basis of unique sets of cultural and religious attributes, such as Jewish Americans" (National Council for the Social Studies 1991). *Religion* alone would not suffice in expressing what defines Jewish American identity. The distinction is seen in the contrast of the word *Judaism*, referring to the religion, and *Jewishness* (or *Yiddishkeit*), referring to that wider set of attributes. *Yiddishkeit* refers not so much to religious commitment as to what the website of the Union of Orthodox Jewish Congregations of America calls "emotional attachment" and "a feeling of identification with the Jewish People" (Orthodox Union 2003). One method for expressing such identification is through dialect. Consciously or subconsciously, group members share distinctive vocabulary, pronunciation, syntax, and discourse styles. An *ethnolect* emerges that insiders use to communicate with each other and that outsiders recognize as a defining group characteristic. Attitudes toward JE, like attitudes toward ethnolects in general, have depended essentially on attitudes toward the group. When the group is viewed with disfavor, then descriptions of the way group members speak are also likely to be unfavorable.

The origins of JE lie in the characteristics that define Jewish Americans as an ethnic group. The mechanisms of its development are embedded in criteria for defining an ethnic group provided by the Task Force on Ethnic Studies Curriculum Guidelines of the National Council for the Social Studies (1991):

   a. Its origins precede the creation of a nation-state or are external to the nation-state. In the case of the United States, ethnic groups have distinct pre-United States or extra-United States territorial bases, e.g., immigrant groups and Native Americans.

   b. It is an involuntary group, although individual identification with the group may be optional.

   c. It has an ancestral tradition and its members share a sense of peoplehood and an interdependence of fate.

   d. It has distinguishing value orientations, behavioral patterns, and interests.

   e. Its existence has an influence, in many cases a substantial influence, on the lives of its members.

   f. Membership in the group is influenced both by how members define themselves and by how they are defined by others.

Each of these characteristics is useful not only in defining Jewishness as an ethnic label but also in seeing how that identity came to be represented through language.

ORIGINS EXTERNAL TO THE STATE

Almost wherever they have settled, Jewish groups have brought with them and maintained language varieties distinct from those of the general population. The *Jewish Language Research Website* documents, in addition to Hebrew and Yiddish, Jewish varieties of Aramaic, Arabic, English, French, Greek Iranian, Italian, Persian, Portuguese, Provençal, and Spanish (Benor 2002). Two main varieties of JE emerged in America, originating from two regionally distinct European groups: Sephardim and Ashkenazim. Sephardic Jews, primarily from Spain and Portugal, immigrated to America beginning in the seventeenth century. Echoes of their presence haunt Henry Wadsworth Longfellow's familiar poem, "The Jewish Cemetery at Newport":

> The very names recorded here are strange, Of foreign accent, and of different climes; Alvares and Rivera interchange With Abraham and Jacob of old times. (Longfellow 1890, ll. 13–16)

The Newport, Rhode Island, cemetery dates back to 1677; the nearby synagogue, to 1763.[1] By the time Longfellow wrote his poem in 1852, however, the Jewish population of the city had largely disappeared:

> Closed are the portals of their Synagogue, No Psalms of David now the silence break, No Rabbi reads the ancient Decalogue In the grand dialect the Prophets spake.
>
> Gone are the living, but the dead remain . . . (ll. 21–25)

Another group of Sephardic Jews emigrated from the Ottoman Empire during the late 1800s and early 1900s. In addition to the languages of their native countries, Sephardic immigrants brought with them a language known as Dzhudezmo (or Judezmo) and its literary counterpart Ladino. The linguistic heritage of these groups is represented in the pronunciation of modern Hebrew spoken in Israel; but Sephardic speakers have had less influence on English in the United States, where many assimilated not only among non-Jews but also among the more populous Ashkenazic Jews.[2]

Ashkenazim arrived in two distinct waves. In the early 1830s, most Jewish immigration to the United States was from Western Europe: Germany, Holland, Alsace, Bohemia, Switzerland, and western Hungary. Later in the nineteenth century, there were increasing numbers from Eastern Europe: Russia, Austria-Hungary, Romania, and what was later Poland. Both groups of Ashkenazim, at least at one time, spoke a common language, Yiddish, in addition to the separate national languages of their countries (see Gold 1981).

Although Yiddish relates linguistically most closely to German, it was primarily the Eastern Ashkenazic group that maintained Yiddish in America. One reason, as suggested by David Gold, was that in the period just prior to emigration, many Western European Jews were already using varieties of German, Dutch, French, or Czech, rather than Yiddish, as the primary spoken language within their separate Jewish communities. Although each of these languages retained some Yiddish expressions, Yiddish was disappearing among the Western Ashkenazic even before they arrived in America. In contrast, most Eastern Ashkenazim, who came to the United States between 1881 and 1924, spoke Yiddish, though not always of the same variety. The rival Litvaks (Jews from Lithuania) and Galitzianers (Jews from Galicia, once part of the Austro-Hungarian empire), for example, spoke two different dialects of Yiddish. Still, it is Gold's contention that most Ashkenazim knew Yiddish and that about five million of the six million Jewish casualties of the Holocaust were Yiddish speakers. At one time, in the United States as well as in the old country, Yiddish was spoken among Jews in secular contexts and also as the language of Torah study. Although Hebrew was considered a more learned language, Yiddish translations of scripture and prayer were available. Yiddish appeared in newspapers, plays, songs, and prose fiction. It was used for scholarly writings in education, history, and folklore. Among second and third generations in the United States, however, use of Yiddish began to decline and English became more common, especially among Jews attending public schools. It came to be associated with older generations, or with childish play and low humor. Such associations lasted until the 1970s, when a revival of interest in Jewish studies promoted interest in Yiddish. In the meantime, JE, for many Ashkenazim, had replaced Yiddish as their primary language.

INVOLUNTARY GROUP MEMBERSHIP:
HOW MEMBERS DEFINE THEMSELVES
AND ARE DEFINED BY OTHERS

Although a decision to be Jewish may seem voluntary, there are at least two senses in which it is not. First, among the deeply religious, whether Jewish by birth or by conversion, there is a strong sense of having had no choice but to be Jewish. According to Jewish law, one is born Jewish by virtue of having had a Jewish mother. In this sense, group membership is involuntary, even though there may be wide variation in degree of association with both religious and cultural aspects of Judaism. Among those who convert to Judaism, especially to Orthodox Judaism, the belief is that the belonging is predestined, and the sense of identification may be even stronger than it is among Jews by birth.

The second sense in which the decision may be involuntary is evidenced by Nazi processes of labeling Jews, without regard to individual religious preference. Nazi German definitions of *Jew* became progressively more racial than religious. In a document issued April 11, 1933, a Jew was defined as a person with even one parent or grandparent of the Jewish religion. In September of 1935 came the "Nuremberg Law for the Protection of German Blood and German Honor," according to which Jews were prevented from marrying German citizens; in the companion "Citizenship Law" of November 1935, a person was defined as fully Jewish if descended from three *racially* Jewish grandparents. That definition was later refined to include the categories of "*Mischling* of the first degree" (two Jewish grandparents) and "*Mischling* of the second degree" (one Jewish grandparent).[3] Those who wanted to further an anti-Jewish agenda developed a way of speaking about Jews that would justify the group's aims. As Hitler announced in an address to the Reichstag on January 30, 1939, the "Jewish question" should be resolved with the annihilation of the "Jewish race."[4] Thus, it is not without some cynicism that Rabbi Morris Kertzer points out in his book *What Is a Jew?*, "It might be well, in considering the question of who is a Jew, to examine into the matter of who is asking the question—and why" (Kertzer 1965, 18).

ANCESTRAL TRADITION AND SENSE OF PEOPLEHOOD

Nothing is more important than tradition in defining Jewish identity. Tradition links Jews to their past, provides rituals for daily living and religious celebrations, and defines values that constitute a Jewish view of the world. Rabbi Kertzer captures this spirit:

> In general, this portrayal of the Jewish way of looking at things attempts to convey some of the warmth, the glow and the serenity of Judaism: the enchantment of fine books; the captivating color of Hasidism; the keen insights of the Babylonian rabbis into human relations; the sane, level-headed wisdom of the medieval philosophers, the mirthful spirit of scholars more than sixteen centuries ago; and the abiding sense of compassion that permeates our tradition. (Kertzer 1965, 17)

Literature, music and prayer, philosophy, humor, and sensitivity to others all involve distinctive linguistic means of conveying ethnic identity. Speech acts of storytelling, singing, praying, debating, joking, consoling bind people to each other and to a shared past.

In the United States today, JE refers to a variety that emerged among Yiddish speakers. Although not all Jews in America share that linguistic tradition, it has become the variety typically associated with American Jews. Adaptations of stories originally written in Yiddish have increased public awareness of Jewish tradition and the language associated with it. Sholom Aleichem's (1894–1914) character Tevye the Dairyman gained fame as the title character in the musical *Fiddler on the Roof*. The lyrics of the song "Tradition" (Harnick and Brock 1964, 6–9) from that musical convey the association of language with traditional family structure:

(TEVYE)

Tradition, tradition! Tradition!

Tradition, tradition! Tradition!

(TEVYE & PAPAS)

Who, day and night, must scramble for a living,

Feed a wife and children, say his daily prayers?

And who has the right, as master of the house,

To have the final word at home?

The Papa, the Papa! Tradition.

The Papa, the Papa! Tradition.

(GOLDE & MAMAS)

Who must know the way to make a proper home,

A quiet home, a kosher home?

Who must raise the family and run the home,

So Papa's free to read the holy books?

The Mama, the Mama! Tradition!

The Mama, the Mama! Tradition!

The father's role as a leader is conveyed through his participation in rituals associated with language. As master, he is the one who must "say his daily prayers," and who "has the final word at home." Mother takes care of the home so that he is "free to read the holy books."

Entrenched gender roles are central to Isaac Bashevis Singer's (1962) story "Yentl the Yeshiva Boy." Originally written in Yiddish in 1962, the story's title character is a young woman who secretly learns to study *Talmud* ("commentary on Jewish law") from her devoted father; after his death, she assumes a male identity in order to continue her study at *Yeshiva* ("Jewish school for boys"). In 1977, Singer and Leah Napolin came out with a stage version, which continues to be produced in Yiddish (it played through December 2002 at the Folksbiene Yiddish Theatre in New York); but it was Barbra Streisand's (1983) film version of *Yentl* that brought Yiddish expressions and traditions to a broader, English-speaking audience.

DISTINGUISHING VALUES, BEHAVIORS, INTERESTS: INFLUENCE ON LIVES OF MEMBERS

Individual American Jews vary in the extent to which group membership dictates daily ritual. One may claim group membership without attending synagogue or temple, without keeping kosher, without studying Jewish law or even celebrating Jewish holidays. Among religious Jews, however, every aspect of life is associated with group membership. Because driving on the Sabbath is forbidden, Orthodox Jews live within walking distance of the synagogue, strengthening the sense of community among group members. They delineate a physical boundary, called an *eruv*, outside of which objects are not carried on the Sabbath and high holidays. Linguistic performances of group members—praying, studying, reading, singing, and speaking—serve to reinforce dialect features. Specialized vocabulary, some of which is discussed below, is required to identify items associated with rites of passage, holiday ceremonies, household items, specialized clothing, and other objects and activities. In this performance of religious observance, Sol Steinmetz (1981) sees the origins of JE in America:

> Modern Orthodox Jews maintain an intimate contact with Yiddish, using it as a source of unlimited borrowings to cover every area of their religious and social life. This mixture of Yiddish and English (and increasingly also of Modern Hebrew) which has evolved in this fashion I have called JE, on the model of the names of other ethnic varieties of English such as black English, Puerto Rican English, and Chicano English.

FEATURES OF JEWISH ENGLISH

Jewish English is characterized by specialized features of vocabulary, grammar, pronunciation, and discourse that set it apart from Standard American English. The degree to which speakers use these features depends upon a number of complex factors: extent of religious

observance; ancestry; place of residence; exposure to other speakers of
JE; and the speaking situation, including audience, occasion, and
medium of communication. There is no one inventory of such features. The discussion that follows is compiled from popular, scholarly,
and religious sources, as well as from my own experience.

Vocabulary

Jewish English comprises a vocabulary derived for the most part from
Yiddish and Hebrew. Inventories of the lexicon are readily available.
Leo Rosten's (1967) *The Joys of Yiddish* is among several of his popu-
lar books on the subject that provide not only definitions but also
extensive explanations and examples.[5] *Meshuggenary: Celebrating the
World of Yiddish*, by Stevens et al. (2002), includes a wonderful col-
lection of Yiddish words, proverbs, insults, and blessings.[6] Gene
Bluestein's (1998) *Anglish/Yinglish: Yiddish in American Life and
Literature* is rich with examples from songs, comic strips, novels, and
book reviews. Chaim M. Weiser's (1995) *Frumspeak: The First
Dictionary of Yeshivish* is geared toward the language of religious
Jews, whereas Sol Steinmetz's 2005 *Dictionary of Jewish Usage* is
intended as a more popular guide. On-line glossaries are also abun-
dant and provide additional useful contexts. One has the advantage of
voiced pronunciation along with definitions (The Kosher Nosh
2003). Others include encyclopedias covering all aspects of Jewish life
(Orthodox Union 2003; Rich 2002). All of these sources recognize
that much of JE is transliterated from either Yiddish or Hebrew, so
that English spelling is inconsistent, although some have recently
begun to adhere to guidelines for transliteration publicized by the
YIVO Institute for Jewish research (2003).

The JE lexicon, like other features of JE, ranges from items in the
mainstream of American English to ones that are highly specialized.
Names for popular holidays and celebrations, such as *Chanukah*
[lit., "festival of lights"] and *bar mitzvah* [lit., "son of the command-
ment"] are used popularly without translation. Others may be named
either in English (*Passover*) or in Hebrew (*Pesach*), depending on
speaker and speaking situation. Apparent synonyms in JE may convey
subtle differences in Jewish identity, as exemplified by names for the
place where Jews worship: Reform Jews typically refer to *temple*,
Conservative Jews to *synagogue*, and Orthodox and Chasidic Jews to
shul. Words referring to more obscure holidays and observances are
less familiar outside the religious Jewish community. Steinmetz gath-
ers such expressions used in Jewish English-language publications of

the 1970s. Here is one example of such an extract:

> Tisha B'Av 5733 it was finally my zechus to be in Eretz Yisroel.
> I thought I would be able to daven at the Kosel. . . . Our madrichim,
> however felt otherwise. (*Jewish Observer*, November 1973, p.
> 28)
>
> *Tisha B'Av* "Ninth of Av" (a fast day), *5733* (year of Jewish calendar),
> *zechus* "privilege," *Eretz Yisroel* "Land of Israel," *daven* "pray," *Kosel*
> "Western Wall," *madrichim* "leaders." (Steinmetz 1981, 5)

Terms collected by Steinmetz refer to all aspects of life: marriage
(e.g., *shadchen* "matchmaker"), death (e.g., *ovel* "mourner"), study
(e.g., *limud* "learning"), prayer (e.g., *tallis* "prayer shawl"), and kinship (e.g., *zeide* "grandfather"). Expressions include preventive terms
(e.g., *halevai* "would that it be so"), greetings (e.g., *boruch habo*
"welcome"), expletives (e.g., *yemach shemo* "may his name be blotted
out"), interjections (e.g., *nu* "well, so"). Some terms have both
Hebrew and Yiddish variants that are used interchangeably. The skull-
cap worn by Orthodox and Conservative Jews, for example, may be
referred to either as a *kippah* (Hebrew) or *yarmulka* (Yiddish).

A substantial number of words and expressions originating in JE
are used popularly in English language media. Stevens et al. (2002,
67–93) mark with an asterisk numerous words that they have found in
English dictionaries. They also list the top twelve hits using the
Internet search engine Google.com (Stevens et al. 2002, 17). These
terms are listed in table 4.1 in the order of frequency in which they

Table 4.1 Popular Jewish English terms on Internet and in newspapers

Top Twelve Terms Using Google[a]	Rank (N): Internet Search Using Google[a]	Rank (N): Newspaper Search Using Lexis/Nexis
glitch "slip-up"	1 (232,000)	1 (794)
kosher "legit, ritually clean"	2 (222,000)	3 (346)
bagel "doughnut-shaped roll"	3 (145,000)	2 (368)
maven "expert"	4 (70,000)	4 (267)
yid "Jew"	5 (62,000)	12 (6)
klezmer "Yiddish folk music"	6 (46,600)	7 (110)
mensch "decent person"	7 (42,600)	10 (30)
tush "backside, rear end"	8 (39,500)	11 (25)
schlock "junk"	9 (39,300)	8 (108)
klutz "clumsy person"	10 (39,000)	9 (73)
schmooze "chat, gossip"	11 (38,100)	5 (218)
chutzpah "impudence"	12 (32,000)	6 (202)

[a] Stevens et al. 2002, 17.

appeared using Google. For comparison, the results of a search of major newspapers using Lexis/Nexis for the period September 1–November 1, 2002 are also listed. Both searches include variations in spelling and form (such as plurals and verb forms). The relative frequency for the Internet and for newspapers is similar, except for the word *yid*. *Yid* appears rarely in newspapers, probably because of its pejorative connotations, although a check of the Internet suggests that it is not used pejoratively on most websites. Other words with more than twenty-five newspaper hits in the same two-month period include *tchotchke* "knick-knack" (51), *schmuck* "jerk" (38), *kvetch* "whine" (37), *nebbish* "nonentity" (31), and *kibitz* "to observe, as in a card game, and give unwanted advice"(29).

These words are common not only in websites and newspapers, but also in magazines, literary works, television shows, and films, sometimes with variant forms or spellings. *Tchotche*, literally a child's toy, usually refers to a knick-knack; it is used in this sense on an episode of *Designing for the Sexes* on Home and Garden Television (2003). It may also, especially in the diminutive form *tchotchele*, describe a pretty but spoiled young woman; in this sense it appears in an example quoted in Bluestein (1998, 114) from Philip Roth's *Portnoy's Complaint*: "O you virtuous Jewess, the tables are turned, *tsatskeleh*." JE words used in popular media do not always maintain the connotations of the original. Whereas *kosher* in JE refers to the ritual adherence to dietary laws, in more general American English it may refer more generally to proper behavior. An article in *The Observer* quotes boxer Budd Schulberg as saying, "My father and I, being pretty knowledgeable fans, felt the fight was not exactly kosher" (Hagerty and Elder 2002). In JE, the word *schmuck* is an obscene term, meaning "penis" and analogous to the word *prick* in American English. In general American usage, however, the term is a less vulgar term for an inept individual, as illustrated by this quote from an announcer at an American football game: "The victory was startling for several reasons, most notably because the Colts had just lost three in a row and their four victories were against a bunch of schmucks" (Pittsburgh *Post-Gazette* 2002). The term is so inoffensive, in fact, that cartoonist Charles Schulz had Lucy invent a *schmuckleball* pitch (see Bluestein 1998, 102). Euphemisms *shmoo* and *shmo* (*schmoe*), both terms for an ineffectual person, originate in American JE. Cartoonist Al Capp's likable and self-sacrificing Shmoos inhabited Li'l Abner's Dogpatch until they were hunted into extinction,[7] and "Joe Schmoe" is used unselfconsciously to refer to an ordinary American guy.

GRAMMAR

"Joe Schmoe" is also an example of Yiddish rhyming slang. Yiddish-sounding *s(c)hm-* is a productive word-formation process when rhymed with an English word to suggest playful dismissiveness. The sense of it is captured by the title of Fran Drescher's (2002) book describing her battle with cancer, *Cancer Schmancer*, and by her words on the book's dust jacket: "Dear Friend," she writes, "All I've got to say is, to hell with cancer. Laughing at the crazy things life offers even when it's biting you on the ass." Bridge champion Marty Bergen (1995, 1999, 2002) offers the book *Points Schmoints* (and its sequels), which details a method of hand evaluation emphasizing suit distribution as opposed to the high-card point count system popularized by Charles Goren.[8] "Stocks schmocks, what about mutual funds?" queries one fund's website (Ferris 2002). "Deficit schmeficit: Not a Bush priority" reads a headline published by *USA Today* (Shapiro 2003).

When Yiddish or Hebrew words become part of JE, they may be integrated through the use of English suffixes. Yiddish verbs, for example, typically lose the *–(e)n* Yiddish infinitive and take on English inflections: Yiddish *bentshn* has become *bentsh* "to recite the Grace after Meals"; *dav(e)nen, dav(e)n* "to pray"; *kvetshn, kvetsh* "to complain"; *shlepn, shlep* "to drag"; *shnodern, shnoder* "donate to the synagogue"; *mutshn, mutshe* "torment"; *farbrengen, farbreng* "hold a [Chasidic] get-together." These are conjugated, then, as English verbs: *bentshes, bentshed, bentshing*; shlep, shlepped, shlepping. Plurals sometimes alternate their forms. Some Yiddish nouns, like English, take an *-s* plural inflection; others use *–im* or *–lekh*. In JE, Yiddish *kneidel* "dumpling" may be pluralized either as *kneidels* or as *kneidlekh*; *shtetl* "small town" may be *shtetls* or *shtetlekh*. In Hebrew, masculine nouns typically pluralize *-im* and feminine, *-os(t)*. The word *talis* (Heb. and Yid. "prayer shawl") may be rendered in JE as *talises* or as *taleisim*. The word *kippah* (Heb. "skull cap") may be either *kippahs* or *kippot* (or it may be expressed as the Yiddish *yarmulkes*) (Gold 1981, 289; McArthur 1992; and Steinmetz 1981, 8).

English suffixes are also used to change the part of speech of Yiddish and Hebrew words integrated into JE. *Shlep* "to drag" becomes *shleppy, shleppily, shleppiness, shleppish, shleppishly; frum* "religious" yields *frummies* "religious ones"; *Yeke* "German Jew" produces *Yekish* and *Yekishness* (MacArthur 1992; Steinmetz 1981, 8). In addition, Yiddish derivational suffixes may be attached to English or Hebrew words: Yiddish (from Slavic) *–nik* "ardent practitioner, believer, lover, cultist or devotee" has given American English *beatnik*, *peacenik*, and *no-goodnik* (Gold 1981, 290; Rosten 2001); JE examples include *JDLnik* "supporter of the Jewish Defense League," *aliyanik* "one who emigrates to Israel," *refusenik* "Soviet Jew refused permission to leave USSR," *Chabadnik* "member of Chabad (Chasidic sect)," and *bal teshuvanik* "Jew who has become religious" (Steinmetz 1981, 8). The diminutive suffixes – *chik* and –*el(e)* are common and may even be combined: *boychik*, *boyele*, and *boychikel* (plural *boychiklekh*) are all fond names for "little boy" (Rosten 2001; Steinmetz 1981, 8). Phrases may include various combinations of English and Yiddish or Hebrew words, such as *matzo balls* "dumplings" or *shana tova card* "Jewish New Year's card" (MacArthur 1992). Other JE phrases include combinations with English *say* and *make*; one may *say kaddish* "recite mourners' prayer" or *say yizkor* "recite memorial prayer"; *make kiddush* "recite prayer over wine" or *make (ha)motzi* "recite (the) prayer over bread." Jews also *sit shiva* "observe seven days of mourning" (McArthur 1992; Rosten 2001, 47; and Steinmetz 1981, 8).

Some JE sentences result from direct translation from Yiddish expressions. Rosten (2001, xiv–xv) offers the following examples:

> Get lost!
> You should live so long.
>
> On him it looks good.
> Wear it in good health.
> Who *needs* it?
> Excuse the expression.
> Okay by me.
> I need it like a hole in the head.
> You should live to be a hundred and twenty.
> My son, the physicist.

Another syntactic feature of JE is called *Yiddish Movement*, whereby an adjective, adverb, or noun that would ordinarily appear at the end of a sentence is moved to the beginning and stressed:

> Smart, he isn't.
> Already you're discouraged?
> My son-in-law he wants to be.[9]

As Ellen Prince (1981, 1996, 9) points out, this structure is similar to a general feature of English syntax, known as *topicalization* or *fronting*, but its use in general English is limited to certain conditions, such as definition or self-correction. In JE, the discourse function derives from Yiddish, where the fronted term is presumed to be already in the mind of the hearer. It effect is often to convey sarcasm, scorn, or contempt. The following exchanges from Philip Roth's *Goodbye Columbus* capture the tone of Yiddish Movement:

"You've got clean underwear?"
"I'm washing it at night. I'm okay, Aunt Gladys."
"By hand you can't get it clean."
"It's clean enough. Look, Aunt Gladys, I'm having a wonderful time."
"Shmutz ['dirt'] he lives in and I shouldn't worry!" (Roth 1963, 54; quoted in Prince 1996, 8)

That night, after dinner, I gave Aunt Gladys a kiss and told her she shouldn't work so hard. "In less than a week it's Rosh Hashana and he thinks I should take a vacation. Ten people I'm having. What do you think, a chicken cleans itself?" (Roth 1963, 86; quoted in Prince 1996, 8)

DISCOURSE

Although individual styles differ among Jewish speakers, researchers have emphasized three types of discourse features associated with JE. First, Jewish speech is characterized as being loud and fast. Deborah Tannen (1981) describes New York Jewish conversational style as overlapping, loud, high-pitched, fast-paced, and accompanied by exaggerated gesture. In Laura Z. Hobson's *Gentleman's Agreement*, a novel that explores antisemitism in 1940s America, this feature is exploited when two girls are recognized as being Jewish due to their heavy makeup and "strident" voices (Hobson 1947, 242). Second, Jewish discourse is considered to be argumentative. Deborah Schiffrin's (1981) study describes Jewish speech style as involving sociable argument, in which speakers repeatedly disagree, remain non-aligned, and compete for turns. Third, and above all, Jewish discourse is associated with sometimes self-effacing humor.

Comedy, in fact, is such a rich part of Jewish life that in 1979 it was estimated that Jews made up eighty percent of professional comedians (Epstein 2001, x). Lawrence J. Epstein attributes this to the experience of Jews as immigrants; comedy is a way to counter poverty and discrimination. Ironically, Jewish comedians often adopt personas consistent with anti-Semitic stereotypes: Jack Benny, the ultimate cheapskate; Ed Wynn and Rodney Dangerfield, the fool; Woody

Allen, the neurotic. Jewish humor, according to Epstein, is characterized by wit and wordplay, a style attributable to the importance of language in Jewish culture: "Words form the center of study, of prayer, and of entertainment. The emphasis on language and on the argumentative patterns of Talmudic reasoning provided Jews with a style of thinking" (Epstein 2001, xviii).

Epstein tells the story of the Borscht Belt, a string of Catskill Mountain resorts given their moniker from the beet soup enjoyed by many Russian Jewish immigrants, famous as a Jewish vacation center from the late 1930s through the early 1960s. Among the names he associates with that entertainment circuit are Milton Berle, Fanny Brice, Mel Brooks, George Burns, Carl Reiner, Neil Simon, Red Buttons, Danny Kaye, Judy Holliday, Jackie Mason, Alan King, Henny Youngman, Buddy Hackett, Joan Rivers, Jerry Lewis, Woody Allen, Sid Caesar, and Joey Bishop. A hotel in the Catskills is the setting for Steve Stern's (1999) story "The Wedding Jester." The main character assumes the role of a *tummler* (JE, from the Yiddish "noisemaker," for the combination emcee and entertainer typical of Borscht Belt resorts), who tells jokes typical of the Borscht Belt comedian:

> I ask her how's the champagne and caviar. "The ginger ale was fine," she says, "but the huckleberries tasted from herring." She says she feels chilly, so I tell her, "Close the window, it's cold outside," "Nu," she replies, "if I close the window, will it be warm outside?"
>
> When my father was dying, I asked him if he had any last wishes. "All I want is you should fetch me a nice piece of your mother's coffee cake from the sideboard downstairs." Then I have to tell him what my mama tells me, that it's for after . . . (Stern 1999, 213)

Pronunciation

Pronunciation of JE is most closely associated with New York City. In a study of Jews and non-Jews from the New York City area, C.K. Thomas (1932) finds the following features of pronunciation to be most closely associated with JE: a raising of pitch and excessive exploding of "t" and "d"; a slight lisping of "s" and "z"; exaggerated hissing of "s"; substitution of "th" or "sh" for "s"; pronunciation of a hard "g" sound in "ing" words, so that the "ing" of *singer* sounds like that of *finger* or *Long Island* sounds like "Long Guy Land"; and the occasional substitution of "k" for "g" as in "sink" for *sing*. Thomas finds some features of pronunciation to be common to both

Jews and non-Jews of New York: loss of distinction between "wh" and "w," so that *which* and *witch* sound the same; intrusive "r," as in *idear* for *idea*; and a number of substitutions in vowel sounds. Current research substantiates the maintenance of a JE pronunciation. According to Tom McArthur, editor of *The Oxford Companion to the English Language*:

> Certain features of Eastern Ashkenazic New York City English of the immigrant generations (c. 1880–1940) are still sometimes heard: pronunciation of such words as *circle, nervous, first* as if "soikel," "noivis," "foist," and an intrusive /n/ in words like *carpenter* ("carpentner"), *painter* ("paintner"). (McArthur 1992)[10]

These and other features stem from Yiddish pronunciation, including two that had been observed by Thomas (1932): hard "g" in "ing" words and overaspiration of "t."[11] McArthur (1992) also mentions confusion of the "s" and "z" sounds in forming certain plural endings and certain Yiddish-derived vowel substitutions. Other pronunciation features derived from Yiddish, according to McArthur (1992), include "pitch, amplitude, intonation, voice quality, and rate of speech." One vowel feature of Yiddish has been Americanized: Yiddish words ending in schwa have come to be pronounced with an "ee" sound, so that *pastrame* becomes *pastrami; khale, khali* "Sabbath loaf"; *shmate, shmati* "rag"; *tate, tati* "daddy"; *Sore, Sori* "Sarah"; and *rebe, rebi* "Chasidic leader" (Gold 1981, 289; McArthur 1992).

JEWISH ENGLISH AND THE STRUGGLE FOR IDENTITY

Jewish identity lies in a delicate balance: the desire to assimilate, weighed against the desire to honor and belong to one's heritage. Of the wave of Jewish immigration early in the twentieth century, Epstein writes, "Even at this point, there was a struggle within the American Jewish soul about whether they should embrace their tradition or their new land" (Epstein 2001, 14). He identifies this dilemma as part of both the motivation and the torment of Jewish comedians:

> They wanted American approval, but they deliberately chose not to discard their Jewishness. They hid it, but did not surrender it. These immigrant Jewish comedians developed a "double consciousness," a sense of being Jewish but having to hide it to win approval and a sense of being

American, but not fully so. Such a "double consciousness" in many ways defined American Jewish life and the Jewish comedians who found success in America. (Epstein 2001, 51)

The struggle of Jewish comedians is a microcosm of the struggle of every Jew who identifies with two sometimes nonharmonious identities. As Epstein puts it,

> Jewish comedians became the shock troops of American Jewish assimilation, gaining acceptance decades before the wider Jewish community did. Many of these comedians embraced both Gentile values and Gentile women with great fervor. Others struggled to define their own relationship to the more traditional organized Jewish community. Jewish comedians therefore became among the first to reflect, although in an exaggerated way, the tortured relationship American Jews sometimes had with their religion and its culture. (Epstein 2001, xx–xxi)

Such struggle is not limited, of course, to Jewish American ethnic identity. The notion of "double-consciousness" is associated with the noted black writer W.E.B. DuBois, who writes in 1903:

> It is a peculiar sensation, this double-consciousness, this sense of always looking at one's self through the eyes of others, of measuring one's soul by the tape of a world that looks on in amused contempt and pity. One ever feels his twoness,—an American, a Negro; two souls, two thoughts, two unreconciled strivings; two warring ideals . . . (DuBois 1903, ch. 1)

The changing language of Jews in America, like that of blacks, parallels the roles of their place in society. As Alamin Mazrui (chapter 2) argues in applying "double-consciousness" both to African Americans and to other blacks living in a Eurocentric world, there is one consciousness representing opposition to the dominant group of English speakers and another linking members of the same ethnic group. Ethnic varieties of English at once serve to identify the speaker with the majority, through the use of English, and with the minority, through variation from "Standard" English.

As Eastern European Jews assimilated into American culture, Yiddish language use diminished. Bluestein writes of this association between language and culture:

> The fate of Yiddish is not unlike that of the Jew in America—for both, the generation gap . . . has exhibited some central changes in the relations between the Jews and their ancestral culture. (Bluestein 1998, 136)

Like Epstein (2001), he recognizes the pain of reconciling dual identities. He summarizes this theme in another Jewish writer's novel:

> Roth's main point in *Portnoy's Complaint* is that the American Jew can no longer claim either the exceptionalism of an earlier generation or the nice balance of being both Jewish and American without enduring a good deal of anguish. (Bluestein 1998, 138)

JE is one vehicle for expressing such a double identity. It provides opportunity for double-voicedness, which following Bakhtin (1975), encodes the double-consciousness of its user. It represents a merging of cultures and can signify a range of affiliation with Jewish ethnic identity.

A final example illustrates this range of Jewish identity and how language is used to represent it. Alfred Uhry (1997), in the play *The Last Night of Ballyhoo*, makes use of JE as a means of revealing the Jewish dilemma. At the opening of the play, the Freitags, a Jewish family of German descent living in Atlanta in 1939, are seen decorating a Christmas tree and preparing for Ballyhoo, a Southern Jewish tradition that once gave young Jewish men and women a chance to meet and a place to go on Christmas Eve. Joe Farkas, a young Jewish man from New York and of Eastern European descent, comes to Atlanta to work for Adolph Freitag and arranges to take his niece to the dance. One sign of difference emerges when Adolph's daughter, upon being introduced to Joe, remarks on the foreignness of his name: "I've never met a Farkas before" (Uhry 1997, 19).[12] Another follows shortly afterward when Joe is asked about his vacation plans:

> *Reba [Aldolph's sister-in-law]:* Tell me, Joe, will [*sic*] be going up to your home for Christmas?
>
> *Joe:* No, ma'am. My boss there keeps me hoppin' too much for that. But it's okay. My family doesn't celebrate Christmas.
>
> *Boo [Adolph's sister]:* I see.
>
> *Joe:* I'll be home for Pesach, though.
>
> *Lala:* Pesach?
>
> *Joe:* Passover.
>
> *Boo:* You remember, Lala. That time we went to the seder supper with one of Daddy's business acquaintances. I believe their name was Lipzin. They lived over on Boulevard or somewhere. You were in the sixth grade. It was very interesting.
>
> (Uhry 1997, 23)

Joe, from New York, is comfortable with the JE *Pesach*, whereas the Freitag family, from Germany, knows only the English equivalent.

When Joe fails to show interest in Lala, claiming he has to be at work early in the morning, her mother Boo declares, "Adolph, that kike you hired has no manners" (Uhry 1997, 26). Reba shows similar prejudice toward East European Jews in a discussion with her daughter Sunny. Gossiping about one Jewish family in Atlanta, Reba claims, "Well, they were the other kind. . . . East of the Elbe. . . . And west of it is us and east of it is the other kind" (Uhry 1997, 42). Later, Joe asks Sunny, "Are you people really Jewish?" Sunny responds by saying, "That's all we wanted—to be like everybody else"; she then tells Joe about the time a non-Jewish friend asked her to join her at the country club swimming pool and she was asked to leave because she was Jewish (Uhry 1997, 49–51). Sunny would like to believe that religion does not matter in America, but Joe reminds her that people like Hitler will never let them forget. While Hitler is annihilating the Jews of Europe, German Jews in America are trying to show their superiority over Jews of Eastern European origin. After learning that Joe has invited Sunny to Ballyhoo, the jealous Lala concludes Act I with the words, "You'll see, you'll see what happens when you come crawlin' to Ballyhoo with a pushy New York Yid tryin' to suck up to his boss and I sweep in with someone who belongs there. When I sweep in on the arm of a Louisiana Weil!" (Uhry 1997, 57). What does happen is that Lala's date, Peachy Weil, goads Joe into leaving the dance, in a scene that echoes Sunny's experience at the "restricted" country club; the Standard Club caters to German Jews, whereas the Progressive Club is for "the other kind." At a time when Christians are excluding Jews from their resorts and country clubs, the very issue that Hobson explores in *Gentleman's Agreement*, Jews are inflicting the same bias on fellow Jews. *Joe* finally explodes over the prejudice to which he has been subjected, taking it out on Sunny and switching into Yiddish as he reaffirms his Jewish identity:

> *Joe:* Right! Whyn't you just call me a kike and get it over with?
> *Sunny:* I think it is over with.
> *Joe:* A shaynim donk in pupik. [lit. "a nice thank you in the navel"]
> *Sunny:* I don't know what you're saying.
> *Joe:* Thanks for nothing.
>
> (Uhry 1997, 92)

Using the names *yid* and *kike* with each other—words more typically used to express prejudice and hatred of non-Jews toward Jews, especially during the years around World War II—is a profound sign of the tension between rival groups of Jews and between two characters seeking their individual and shared identity. If language separates them here, it brings them together in the end. Sunny feels the gap that has separated her from her heritage: "There's just a big hole where the Judaism used to be. But I remembered I do know some Yiddish. I went to my suitemate's house in Chestnut Hill for dinner once and they said it at the table. Shabit Shallim—something like that." Joe corrects her: "It's not Yiddish. It's Hebrew. Shabbat Shalom. It's the blessing you say Friday night." In the brief final scene of the play, all of the dialogue is in Hebrew. The family is gathered round the Sabbath dinner table, and the Christmas tree has disappeared. After Sunny lights the candles and recites the complete blessing in Hebrew, each family member repeats the Sabbath greeting, "Shabbat Shalom" (Uhry 1997, 97–99).

Although Uhry's play takes place in 1939, *The Last Night of Ballyhoo* is as relevant for Jews and non-Jews today who find themselves simultaneously part of two cultures and sometimes uncomfortable with both. In fact, it was first produced at the Alliance Theatre in Atlanta, where it marked the city's hosting of the 1996 Olympics, and it opened on Broadway the following year. The story of finding one's identity in the American melting pot, of combating prejudice and self-hatred, and of searching for identity through language is one that can be shared with and appreciated by audiences of all ethnic backgrounds.

Other exhibits of Jewish ethnic identity are evidence of its presence in American culture today. Jewish comedians still frequent the airwaves, their styles ranging from the loud and confrontational Howard Stern to the quiet and mild Jerry Seinfeld. Stereotypes of Jewish women, such as the *Yiddishe Mama* and the *Jewish American Princess*, are reflected in the comic antics of television characters Roseanne in *Roseanne* (1988–1997) and Fran Drescher in *The Nanny* (1993–1999). Films such as *Life Is Beautiful* (1998), *The Pianist* (2002), and *Nowhere in Africa* (2003) tell stories of Jewish survival during Holocaust Germany. A revival of klezmer bands, begun in the 1970s, has given Jewish secular music a place among popular varieties of musical styles. All of these have increased public awareness of Jewish language and how it is used to represent Jewish identity.

Language is but one means of representing identity, but it an important one. JE serves to unify Jews with a shared history and culture. At the same time, it distinguishes among Jews according to the countries from which their families came, the extent to which they participate in traditional religious observances, and the extent to which they identify with Jewish culture. In short, it expresses one's view of the world. It represents the struggle of individuals to connect simultaneously with long-standing ethnic traditions and with mainstream contemporary American culture. In the process, JE has become part of that culture. Through print media, theater, film, music, and the Internet, people of all ethnic backgrounds share in JE words, sounds, sentences, and styles that have become part of American language.

Notes

1. For additional information on the Jewish families, synagogue, and cemetery of Newport, see Stokes and Stokes (2003).
2. For further information on Sephardic history, culture, and language, see American Sephardi Association (2003); and Sol Levenson (1990).
3. For further discussion of these events, see Stein (2002).
4. For text and audio of Hitler's speech, see Hornshoj-Moller (1998).
5. Rosten came out with *Hooray for Yiddish: A Book about English* in 1982 and *The Joys of Yinglish* in 1990. *The Joys of Yiddish* was revised by Lawrence Bush and published as *The New Joys of Yiddish* in 2001.
6. *Meshug[g]a* and *meshug[g]ana* mean "crazy"; *meshuggenary* is an invention of the authors.
7. For an informative website on Al Capp's *Li'l Abner*, with particular information about the Schmoos, see Capp Eterprises (2003).
8. Also published by the author are *More Points Schmoints!* (1999) and *Hand Evaluation: Points Schmoints!* (2002).
9. The first two examples are from Prince (1996). The third is from Rosten (2001, xv), where it appears as an example of what Rosten calls "mordant syntax," which is not limited to the fronting process.
10. Interestingly, Gold (1981, 289) considers these two Yiddish-based features to be disappearing.
11. On pronunciation of /t/, see Benor (2001).
12. *Farkas* is Hungarian, meaning "wolf." *Freitag* is German, meaning "Friday."

References

Aleichem, Sholem. 1996 [1894–1914]. *Tevye the Dairyman and The Railroad Stories*, Hillel Halkin (trans.). New York: Schocken Books, Division of Random House.

American Sephardi Association. 2003. *Sephardic House*. http://www.sephardichouse.org/index1.html. Website last accessed May 15, 2003.

Bakhtin, Mikhail M. 1975 [1981]. *The Dialogic Imagination: Four Essays*, Michael Holquist (trans.). Carl Emerson and Michael Holquist (eds.). Austin: University of Texas Press.

Benor, Sarah Bunin. 2001. The learned /t/: Phonological variation in ortho-
    dox Jewish English. *Penn Working Papers in Linguistics: Selected Papers
    from NWAV 29*, Vol. 7.3.
Benor, Sarah Bunin. 2002. *Jewish English Research Website.* http://www.
    jewish-languages.org/languages.html. Last modified December 27, 2002.
    Website last accessed May 15, 2003.
Bergen, Marty. 1995. *Points Schmoints! Bergen's Winning Bridge Secrets.*
    New York: Bergen Books.
Bergen, Marty. 1999. *More Points Schmoints!* New York: Bergen Books.
Bergen, Marty. 2002. *Hand Evaluation: Points Schmoints!* New York: Bergen
    Books.
Bluestein, Gene. 1998. *Anglish/Yinglish: Yiddish in American Life and
    Literature*, 2nd ed. Lincoln: University of Nebraska Press.
Capp Enterprises. 2003. *Shmoo.* http://www.lil-abner.com/shmoo.html.
    Website last accessed May 15, 2003.
Drescher, Fran. 2002. *Cancer Schmancer.* New York: Warner Books.
Du Bois, W.E.B. 1903 [1999]. *The Souls of Black Folk: Essays and Sketches.*
    Chicago: A.C. McClug & Co.; Cambridge, MA: University Press John
    Wilson and Son, 1903; Bartleby.com, 1999. www.bartleby.com/114/.
    Last accessed May 15, 2003.
Epstein, Lawrence J. 2001. *The World of Jewish Comedians.* Cambridge, MA:
    Perseus Books Group.
Ferris, Jonas Max. 2002. So sue them: Shareholder lawsuits on the rise.
    Maxfunds.com, inc. April 3, 2002. http://www.maxfunds.com/
    content/ii040302.html. Website last accessed May 15, 2003.
Gold, David L. 1981. The speech and writing of Jews. In *Language in the
    USA*, Charles A. Ferguson and Shirley Brice Heath (eds.). Cambridge:
    Cambridge University Press. 273–292.
Hagerty, Bill and Neville Elder. 2002. Budd the wiser. *The Observer.*
    December 1, 2002, 55. LexisNexis: http://exlibris.lib.memphis.edu:
    2056/universe/. Website last accessed May 15, 2003.
Harnick, Sheldon (lyrics), and Jerry Brock (music). 1993 [1964]. Tradition.
    In *Songs from Fiddler on the Roof.* Warner Brothers Publications. 6–9.
Hobson, Laura Z. 1947. *Gentleman's Agreement.* New York: Simon and Shuster.
Home and Garden Television. 2003. Family room fix-up. *Designing for the
    Sexes*, DSX-104, March 1.
Hornshoj-Moller, Stig. 1998. http://www.holocaust-history.org/der-ewige-
    jude/hitler-19390130.shtml. Last modified October 24, 1998. Website
    last accessed May 15, 2003.
Kertzer, Morris J. 1965. *What Is a Jew?* 2nd ed. New York: MacMillan.
The Kosher Nosh, Inc. 2003. The unequivalent table of equivalents.
    http://www.koshernosh.com/dictiona.htm. Last modified January 24,
    2003. Website last accessed May 15, 2003.
Levenson, Sol. 1990. *Wandering Thoughts on the Sephardim and Their Language,
    Ladino.* http://www.dartmouth.edu/~library/Library_ Bulletin/Apr1990/
    LB-A90-Levenson.html. Website last accessed May 15, 2003.

LexisNexis   Academic.   http://exlibris.lib.memphis.edu:2056/universe.
Website last accessed May 15, 2003.
Life Is Beautiful (La Vita E Bella). 1998. Directed by and starring Robert
Benigni. Produced by Miramax.
Longfellow, Henry Wadsworth. 1890. The Jewish cemetery at Newport. In
*The Poetical Works of Henry Wadsworth Longfellow, with Bibliographical
and Critical Notes*, Riverside ed. (Boston and New York: Houghton,
Mifflin), III, 33–36. Available at *Representative Poetry Online*,
I. Lancashire (ed.), University of Toronto. http://eir.library.utoronto.ca/
rpo/display/poem1328.html. Website last accessed May 15, 2003.
McArthur, Tom, ed. 1992. Jewish English. *The Oxford Companion to the
English Language*. London and New York: Oxford University Press.
http://www.xrefer.com/Jewish English short form JE.htm. Last accessed
April 19, 2002.
The Nanny. 1993–1999. Prod. by Robert Sternin, Prudence Fraser, Peter
Marc Jacobson, Fran Drescher, and Frank Lombardi. Starring Fran
Drescher and Charles Shaughnessy. Sony Pictures Television.
National Council for the Social Studies. 1991. NCSS Task Force on Ethnic
Studies Curriculum Guidelines, adopted 1976, rev. 1991. http://
databank.ncss.org/article.php?story=20020731121719475. Website last
accessed May 15, 2003.
Nowhere in Africa (Nirgendwo in Afrika). 2003. Directed by Caroline Link.
Starring Juliane Kohler and Merab Ninidze. Produced by Zeitgeist Films.
Orthodox Union. 2003. *Judaism 101*, Glossary. http://www.ou.org/about/
judaism/index.htm. Website last accessed May 15, 2003.
The Pianist. 2002. Directed by Roman Polanski. Starring Adrien Brody.
Produced by Studio Canal.
Pittsburgh *Post-Gazette*. 2003. November 24, D-4, LexisNexis: http://
exlibris.lib.memphis.edu:2056/universe/. Website last accessed May 15,
2003.
Prince, Ellen F. 1981. Topicalization, focus-movement, and Yiddish-movement:
A pragmatic differentiation. In Proceedings of the Seventh Annual Meeting
of the Berkeley Linguistics Society, D. Alford et al. (eds.). 249–264.
Prince, Ellen F. 1996. Constructions and the syntax–discourse interface.
Ms. available at http://www.ling.upenn.edu/~ellen/home.html. Last
accessed April 11, 2003.
Rich, Tracey R. 2002. Glossary of Jewish terminology. *Judaism 101*.
http://www.jewfaq.org/glossary.htm. Last updated November 25, 2002.
Website last accessed May 15, 2003.
Roseanne. 1988–1997. Directed by Ellen Falcon et al. Produced by Carsey-
Warner Company. Starring Roseanne [Arnold / Barr] and John
Goodman. ABC.
Rosten, Leo. 1967. *The Joys of Yiddish*. New York: McGraw-Hill.
Rosten, Leo. 1982. *Hooray for Yiddish: A Book about English*. Simon and
Schuster.
Rosten, Leo. 1990. *The Joys of Yinglish*. New York: Penguin Books.

Rosten, Leo. 2001. *The New Joys of Yiddish*. Rev. Lawrence Bush. New York: Random House.

Roth, Philip. 1959. *Goodbye Columbus*. Boston: Houghton Mifflin; reprinted New York: Bantam, 1963.

Schiffrin, Deborah. 1981. Jewish argument as sociability. *Language in Society* 13: 311–335.

Shapiro, Walter. 2003. Deficit schmeficit: Not a Bush Priority. *USA Today Online*, February 5. http://www.usatoday.com/usatonline/20030205/4838768s.htm. Website last accessed May 15, 2003.

Singer, Isaac Bashevis. 1983 [1962]. *Yentl, The Yeshiva Boy*, Marion Magid and Elizabeth Pollet (eds.). New York: Farrar, Straus and Giroux.

Singer, Isaac Bashevis and Leah Napolin. 1977. *Yentl*. New York: Samuel French, Inc.

Stein, S.D. 2002. Destruction of European Jewry explanatory timeline. http://www.ess.uwe.ac.uk/genocide/destrtim.htm. Last modified September 18, 2002. Website last accessed May 15, 2003.

Stokes, Theresa and Keith Stokes. 2003. Jewish History. In *Eyes of Glory: Two Hundred Years of Ethnic American History*. http://www.eyesofglory.com/jewhist.htm. Website last accessed May 15, 2003.

Steinmetz, Sol. 1981. Jewish English in the United States. *American Speech* 56: 3–16.

Steinmetz, Sol. 2005. *Dictionary of Jewish Usage: A Popular Guide to the Use of Jewish Terms*. Lanham, MD: Rowman & Littlefield.

Stern, Steve. 1999. *The Wedding Jester*. St. Paul, MN: Gray Wolf Press.

Stevens, Payson R., Charles M. Levine, and Sol Steinmetz. 2002. *Meshuggenary: Celebrating the World of Yiddish*. New York: Simon and Schuster.

Streisand, Barbra, dir. 1983. *Yentl*. Starring Barbra Streisand, Mandy Patinkin, and Amy Irving. United Artists.

Tannen, Deborah. 1981. New York Jewish conversational style. *International Journal of the Sociology of Language* 30: 133–149.

Thomas, C.K. 1932. Jewish dialect and New York dialect. *American Speech* 7: 321–326.

Uhry, Alfred. 1997. *The Last Night of Ballyhoo*. New York: Theatre Communications Group.

Weiser, Chaim M. 1995. *Frumspeak: The First Dictionary of Yeshivish*. Northvale, NJ: Jason Aronson.

YIVO Institute for Jewish Research. 2003. *What Is Yiddish? Alef-beys (Yiddish Alphabet) & Transliteration Chart*. http://www.yivoinstitute.org/yiddish/alefbeys_fr.htm. Website last accessed May 15, 2003.

5

LINGUISTIC DISPLAYS OF IDENTITY AMONG DOMINICANS IN NATIONAL AND DIASPORIC SETTINGS*

Almeida Jacqueline Toribio

INTRODUCTION: SITUATING THE STUDY OF DOMINICAN SPEECH PRACTICES

The present work, one instalment of a larger research venture, is motivated by an interest in the extent to which Dominican ethnic identity may be mediated or ascribed via linguistic attributes in national and U.S. diasporic settings.[1] As a study of Dominican speech practices, the endeavor is concerned primarily with aspects of the discipline of sociolinguistics broadly construed. However, its focus is different from that of other studies that appeal to the functional distribution of speech forms.[2] In particular, this study departs from those whose emphasis is in analyzing social structures by appeal to quantifiable linguistic data (cf. the works of W. Labov 1972; L. Milroy 1980, 1987; R. Macaulay 1975, 1991; D. Preston; P. Trudgill 1974, 1983, 1986). It does not present replicable models or statistical evidence of unique language behavior among Dominicans, but attends instead to the careful and deliberate description of language variation and language use as social phenomena (cf. the works of N. Coupland; J. Gee; H. Giles 1977, 2002; J. Gumperz 1972, 1982; B. Rampton 1995). Thus, in tandem with research and theory in communication science and the sociology of language, this work explores the ways in which interactants project identities and define social relations through language performance.[3] In drawing into focus the situated salience of Dominicans' identities, this work seeks to unearth the ecology and economy of code choice implicated in the communicative interactions and social contexts in which Dominicans find themselves (cf. the works of J. Irvine 1989; S. Gal 1995, 1998; K. Woolard 1985).[4] So conceived, this properly linguistic inquiry into Dominican speech practices is consonant with the initiatives of social science scholars in Dominican Studies.

Historian and cultural critic S. Torres-Saillant relates the scholarly study of matters Dominican to the emergence in the 1960s of the larger sphere of ethnic studies (Torres-Saillant 2000).[5] Owing to the labors of an ever-increasing cadre of researchers in economics, sociology, political science, and history, the last decades have witnessed significant production and promotion of knowledge of the Dominican experience abroad—addressing, among others, the causes of migration among Dominicans, their demographic characteristics, their migratory patterns, their mode of incorporation into U.S. society, and their impact on home and host countries (cf. Duany 1994; Graham 1998, 2001; Itzigsohn et al. 1999).[6] Nevertheless, little is known of the language situation of Dominicans in the diaspora (cf. the works of B. Bailey for a notable exception); this in marked contraposition to the vast body of linguistic literature that has profiled the language situations of other prominent Hispanic ethnic groups.[7] Unaddressed until recently were important themes that occupy the language sciences, themes surrounding language loyalties and ethnic boundedness, that engage closely with the core concerns that had largely delimited the ambit of Dominican Studies. Speaking pointedly to Dominican diasporic settlements in the United States, questions regarding the social commentary inherent in codified linguistic gestures loomed large: What roles are accorded to heritage and dominant language in identity formation and community building? What dimensions of the host society are at play in the activation of selected language displays?

This chapter, like previous works by the present author, contributes to redressing the aforementioned oversights and lacunae in linguistic and Dominican scholarship through a broad examination of Dominican language use in the homeland and in the diaspora. The language data to be scrutinized are drawn from a sampling of interviews with participants representative of both sexes and diverse ages and socioeconomic classes in the Dominican Republic and in the New York metropolitan area over the past five years. The main method of data elicitation is a modified interview technique, informed by the insights of ethnographers: all informants were invited to participate in a semistructured discussion on Dominican cultural traditions and societal norms with a known Dominican investigator. A guiding set of questions encompassed three broad areas: personal information, indicators of linguistic insecurity, and perspectives on Dominican ethnicity in the Caribbean and abroad. The individual sessions were recorded and subsequently analyzed and interpreted with an eye toward assessing speakers' exploitation of linguistic forms in the communication of social meaning.

The kernel of the essay is organized as follows. In the second section, Dominicans are revealed to deploy a stigmatized variety (vis-à-vis Peninsular Spanish) of a stigmatized language (vis-à-vis English) in binding themselves to their Dominican compatriots and isolating themselves from their African and African American neighbors. In the third section it is shown that when ethnic identity is perceived as important, Dominicans will make themselves favorably distinct on dimensions such as language (cf. proposals based in Giles's ethnolinguistic identity theory). In the fourth section, it is argued that the sociocultural context that frames discursive events may act as a constraint on the social meanings that are available to be constructed and inferred from linguistic acts (cf. the "acts of identity" framework of Coupland 2001; Goffman 1974; Le Page and Tabouret-Keller 1985, chapter 1 in this volume). In the fifth section, Dominicans are observed to select languages and language varieties in anticipating and serving oftentimes contradictory social and identity outcomes (cf. Bell's 1984 "audience design"). Finally, the sixth section presents compelling evidence of the emergence of new identificational discourses fashioned by second-generation adolescent U.S. Dominicans. The work concludes with summary remarks and recommendations for further research in sociolinguistics and cognate areas.

LINGUISTIC ACTS OF IDENTITY ON THE ISLANDS OF MANHATTAN AND HISPANIOLA

In New York City and the surrounding area, Dominican Spanish is identified and evaluated as being of marginal status (cf. García et al. 1988; Pita and Utakis 2002; Toribio 2001; Zentella 1997).[8] On introspection, Dominicans characterize their speech as provincial, whereas they describe other Spanish varieties as merely different:[9]

(1) a. Dominicans don't speak Spanish well. I'm not saying that I speak perfect Spanish or perfect English . . . All you see is Dominicans that are from *el campo*. Everybody knows right away that they're Dominicans; you get embarrassed because of those people. (Gina, NY working-class female, age 30)

b. [Translation] If you spent a day at my job, you would notice that my form of speaking is a mix of all of the different types of countries' races. The problem is that where I work is a pharmacy and I am there more because I can speak Spanish . . . And people tell me, "But you don't speak like Dominicans." (María, NY working-class female, age 24)

These insecurities are harbored even in the Dominican Republic, where speakers are not confronted with the direct criticism from (or comparison to) speakers of other Spanish dialects. The Dominican vernacular is stigmatized and aesthetically undervalued for lacking certain features of an idealized standard—the Castilian or "European" variety:[10]

(2) a. [Translation] I like the way the Spaniards speak . . . they have better form than us speaking. (Santo, DR middle-class male, age 30)
   b. [Translation] The Spanish of Spain is more refined. The Spanish language came from Spain, didn't it? (Chato, DR middle-class male, age 54)

Especially salient to the native (and nonnative) ear are linguistic variations based in regional pronunciation.[11] To be sure, social, economic, and secular rivalries also separate regions of the Dominican Republic, thereby reducing interregional mobility and augmenting regional idiosyncrasies. But Dominicans' dim view of the northwestern variety, where the contact with the neighboring nation, Haiti, is most pronounced, speaks, perhaps less obliquely, to the discounting or disparaging of the Haitian Creole, as confessed in (3).

(3) a. [Translation] We speak more-or-less regular here . . . The region that speaks poorly, that speaks somewhat tongue-tied, is in Vaca Gorda, because there they are all blacks. It's as though their tongues are tied, they are somewhat Haitianized. They are here as if they were Dominicans. (Donaldo, DR working-class male, age 70+)
   b. [Translation] Here in El Rodeo there was some Haitian heritage; in that area of El Rodeo people didn't speak Spanish well. They sometimes used dialectical words, that sometimes we ourselves didn't understand, that same class of people, Haitians, who mixed in there. (Chato, DR middle-class male, age 54)

The popular view in the Caribbean nation is that the "best" Spanish variety approximates the Castilian norm, and the "worst" variety is spoken by those who, by dint of birth or social circumstance, are believed to be influenced by the African substratum (cf. Zentella 1997 for discussion of attitudes toward Caribbean Spanish in general and Dominicans in particular). In this predilection for the northern peninsular variety and emphatic repudiation of the influence of the Haitian Creole, Dominicans make a great deal of their *hispanidad* while at once racializing the Haitians, in a code of belief and conduct to which we briefly digress. This cursory overview of the backdrop of racial classifications that operate in the Dominican Republic will prove vital to understanding how Dominicans manage their identities in the United States.

Throughout its history, the Dominican Republic has held an unofficial policy against *negritude*, and an official policy of affirmation of the nation's Spanish roots—recall the foregoing discussion of Dominicans' privileging of Peninsular Spanish. The result has been a propagating of the sentiment that African heritage is negative and shameful, and an enforcing of European supremacy, positions that Dominicans publicly uphold (cf. Baud 1997; Cambeira 1997; Sorensen 1997). The popular anti-Haitian position was given substantive and highly animated expression by most of the Dominicans interviewed for the present project. Some believed that this stance dated to the war of independence against Haiti, and to the subsequent Haitian occupation of Santo Domingo in the early nineteenth century. For others, more recent memory invokes the Trujillo dictatorship, which shaped Dominican racial attitudes in profound ways. In promoting his doctrine of *hispanidad*, which defined Dominicans as the purest Spanish people in the Americas, Trujillo put forth a number of maneuvers to deliver the Dominican nation from "Haitianization," employing a simple linguistic litmus to sort friend from foe. The "offending" Haitians were to be identified by their inability to offer a native Dominican pronunciation of the word *perejil* ("parsley"), the assumption being that the uvular trill of Creole speech would compromise a speaker's Haitian identity. (This is reminiscent of the biblical passage of the Gileads' identification of comrades through the pronunciation of the word "shibboleth.") Ongoing border patrol detention drills are keenly reminiscent of these tactics:

(4) [Translation] Your uncle Otilio was in Dajabón, picking up Haitians for immigration. Then, the truck was full, en route to take them to Haiti, and when they were going towards the border, there was a dark man sitting in the park. Otilio says, "Let me see, let me check that dark man, to see." Otilio got out and says to him, "Come here.

Are you Dominican?" And the Haitian says, "Yes!" Otilio said, "If you are Dominican, you will repeat what I tell you." The Haitian says, "Alright." Otilio says, "Repeat this: General Rafael Leónidas Trujillo, benefactor of the new motherland, born in San Cristóbal, the town of *perejil.*" The Haitian says, "Why not just tell me to get in the truck." (Domingo, NY working-class male, 45)

The national "othering" of Haitians has proven so effective that many Dominicans continue to believe that Haitians are the only blacks on the island of Hispaniola. Today, light-skinned Dominicans and darker-skinned members of the middle and upper classes call themselves white, whereas the vast majority call themselves *mestizo* or *mulatto,* though even within these categories, numerous subtle shadings are recognized—for example, *trigueño, indio claro, indio oscuro, canela, moreno* (all forms produced in the interviews).[12] This classification in mind, Dominican migrants abruptly apprehend that the United States is not just racist, but color-blind with regard to gradations of black and white (cf. Landale and Oropesa 2002; Levitt 2001; Rodríguez 2000). In the dualistic black/white racial system, it doesn't matter what they believe they are: objectively speaking, in the United States, Dominicans are African descendants (cf. Baud 1997; Duany 1994; Grasmuck and Pessar 1996; Moya Pons 1981). Naturally, they are often ill-prepared to interpret and accept discrimination on these grounds:[13]

(5) a. We don't consider ourselves black and we don't consider ourselves white. White people don't consider us white, we're like peach. And the black people consider them [Dominicans] brown, so Dominicans are between black, brown, and peach. (Miguelito, NY working-class male, age 11)
b. [Translation] For the white we fall in the black . . . that is for the whites. The white person doesn't distinguish between the light and the black, instead s/he conceptualizes it all in the same frame. (Quiño, NY working-class male, age 60+)

As a first strategy for navigating race relations in the United States, Dominican immigrants make themselves immediately distinct from African Americans (cf. Levitt 2001),[14] and, logically, language affords one simple means of doing so:

(6) a. . . . sure, you're Hispanic, but you're considered black. . . . When you talk, they can tell. (Gina, NY working-class female, age 30)
b. [Translation] In speech you know. There are black Cubans and from other countries. (Quiño, NY working-class male, age 60+)

c. I thought she was African American until she started to talk Spanish . . . (Manuel, middle-class male, age 13)

Thus, not unlike the *perejil* touchstone of the Trujillo era, the Spanish language plays a vital separatist function, isolating Dominicans from their African and African American neighbors. On the island, the Spanish language serves as a marker of a national/cultural status—Dominican, therefore non-Haitian, that is, non-black; or *Spanish*, therefore European, that is, non-African. In similar fashion, in the United States the heritage language is believed to serve as a marker of *ethnic* grouping—immigrant and therefore exempted from the dualistic black/white classification.

To recapitulate, Dominican immigrants' awareness of speaking a stigmatized language (relative to English in the United States) is in competition with the perpetuation of social structures. Interpreting the introspections proffered in the larger project, it is suggested that the reality of the limitations imposed by racial ideologies in the United States may have important consequences in the linguistic behaviors of the New York Dominicans studied. Although their heritage language forms are readily identified and recognized as being of low prestige, New York Dominicans would accrue little benefit in acquiring a pan-Hispanic norm or relinquishing their native language in favor of English. Indeed, the deficit would be dual for the Dominican immigrant who relinquishes native linguistic (and cultural) ties and nonetheless remains the object of discrimination.

The Salience of Ethnolinguistic Characteristics

As shown, although identity is determined by a multiplicity of contributing factors (e.g., language, religion, race, national origin), the social contexts pertinent to Dominicans accredit language a central role: language is recognized as conveying information about their status, and is knowingly exploited in the display of ethnic identity. Ethnic identity refers to a subjective experience that comprises self-perception, a sense of shared values and feelings of belonging:

> An "ethnic group" is a reference group invoked by people who share a common historical style (which may be only assumed), based on overt features and values, and who, through the process of interaction with others, identify themselves as sharing that style. "Ethnic identity" is the sum total of feelings on the part of group members about those values,

symbols, and common histories that identify them as a distinct group "Ethnicity" is simply ethnic-based action. (Royce 1982, 18)

U.S. Dominicans deem language to be a crucial aspect of their identity, a positive assertion and enactment of *dominicanidad* (cf. proposals based in Giles's ethnolinguistic identity theory):

(7) a. [Translation] Dominican culture comprises language. I would say that to be Dominican and to speak [Spanish] is important, not to say original/characteristic. Dominicans who don't speak [Dominican] can feel equally proud, but they are lacking something. (María, NY working-class female, age 24)
   b. [Translation] Dominican without Spanish? That's something you carry in your heart. But for you to say that you are Dominican, you have to speak Spanish. (Pedro, NY middle-class male, age 33)

Some informants are more explicit still in voicing the isomorphism between language and cultural/ethnic identity; observe the reports of intercultural exchanges with interactants of Italian heritage:

(8) a. [Translation] As I tell my Italian friends. How can you say you're Italian if you don't speak the language? (Pedro, NY middle-class male, age 33)
   b. I took Manuel with me to the game and we're practicing, and there's a shortstop and he goes, he asked Manuel whether he speaks Spanish and Manuel said no. And I said, "Manuel don't lie," because I was speaking Spanish to him. And then he [the shortstop] made a comment, and I said, "Well, what do you categorize yourself as?" Like, "Italian?" and I said, "Well do you speak Italian?" He said no. I said, "Well I'm Spanish: ¿Cómo tú estás? Yo me siento muy bien. Mi nombre es Felipe." He got so pissed. . . . You know, he's got a little nice Infinity [automobile], he's got some nice rims, he's got a nice little Italian sticker on the car, Italian things inside the car. "You don't know Italian." I got my Dominican flag in there, I'm Dominican. I know Spanish. (Felipe, age 32)

The above statements speak to the core value attributed to the heritage language as a feature of group membership, as a cue for ethnic categorization, and as a means of ingroup cohesion (cf. Giles and Coupland 1991). (Observe in this respect that the informants quoted in (7–8) are of white phenotype and draw on Spanish primarily in its unifying rather than separatist function.) Thus as an external behavior language allows for the identification of a speaker as a member of some group; but it also permits a means of identifying oneself, as included in or excluded from a particular grouping (Tabouret-Keller 1997). This linguistic variability may be glossed within Le Page and Tabouret-Keller's (1985, 81) "acts of identity" framework, according to which

> [T]he individual creates for himself the patterns of his linguistic behaviour so as to resemble those of the group or groups with which from time to time he wishes to be identified, or so as to be unlike those from whom he wishes to be distinguished.

Of course, the extent to which Dominicans accentuate their linguistic features—thereby insisting on the development of an ethnolinguistic identity—depends on the interplay of a number of sociological, sociolinguistic, and psychological factors. Recall that in the "acts of identity" orientation, language is conceptualized as being imbued with social meanings, that is, sociocultural associations and implications (cf. Bakhtin 1981; Gumperz 1972 on "heteroglossia"). However, the dictum that underlies this philosophy toward the fulfillment of communicative achievements merits a measure of probing.

THE FRAMING OF
LINGUISTIC PERFORMANCES

Though it is true that speech can both produce and occlude symbolic meanings, Le Page and Tabouret-Keller's viewpoint would grant U.S. Dominicans a degree of self-determination that their experiences belie. This criticism is made explicit in Coupland (chapter 1):

> [P]articular discursive frames posit specific affordances and constraints for interactants at specific moments of their involvement, foregrounding certain types of identity work that *can* [italics mine] be done at those moments, and either giving relevance to or denying relevance to certain categories of linguistic indexicals.

Following Coupland, in "doing identity work" speakers position themselves (and hence, others) in relation to sociocultural and sociopolitical community arrangements (what he terms the "sociolinguistic ecology" of the community, i.e., the linguistic resources that are made available by the structure of the community, the sociopolitical value-systems that these resources index). However, Coupland's standpoint may itself be insufficiently restrictive, as the sociolinguistic framing of a speech event oftentimes dictates the types of identity work that *must* be done. A ready illustration of the need for elaboration or clarification of Coupland's proposition is found in examining the language behaviors of black versus white Dominicans in New York. As recognized in (9), the context of the U.S. society and its attendant racial ideologies do not simply permit but clearly *command* a determined linguistic performance of black members of the Dominican diaspora.

(9) a. People, like, ask me if you know English and Spanish . . . I say, like, "Yeah," and then, like, one of the Spanish kids come, and then, like, I have to talk Spanish . . . and then if they say, "Yes," that means I do know it, and if they say, "No," it means I don't speak Spanish. [They are testing me] because they can know me well . . . They always find out [I'm not African American] because they ask me weird questions, like, how old am I [in Spanish]. I get very confused and I forget everything. (José, NY middle-class male, age 8)[15]

b. Well, it [the "trial"] usually happens like once a month. Because every month I meet somebody new and then they're like, "Where are you from?" And I ask him like, "What are you?" They say, "I'm plain American." And then they ask me, "What are you?" I'm not plain American. I'm American and I'm Dominican . . . There's this kid in my school who doesn't believe a lot, because he always asks kids, "Where are you from?" They usually lie, like, "I'm from Puerto Rico." And he's like, "Then you can speak Spanish." And they're like, "No." And he's thinking that they're lying even if they really are. (José, NY middle-class male, age 10)

What is pertinent for the present discussion is not that the boy quoted in (9), in two interview sessions, possesses the linguistic resources to elide his blackness, but that their deployment is compulsory and definitive in defined milieus.

The issue of the agency that is or is not afforded to Dominican interactants in New York communities is further evinced in considering the differential roles scripted for Dominicans of fair appearance. Their assimilation into the mainstream society facilitated, phenotypically White Dominicans often become promoters of a personal narrative and political discourse that echo those of the European American majority (Toribio 2003).

(10) [Progress] depends on the individual. This [the U.S.] is where you can set your goals and accomplish whatever you want. . . . [African Americans] feel they're minorities same as us, so we have to try to team up . . . [an alliance] . . . but it's more speculation than anything else . . . You would run into an African American and he's into the old times when there were slaves and he feels that everyone is against him. And then you run into someone, like, "Hey, that didn't happen to me. . . ." And also you feel that you run into African Americans who all they want to do is play the race card game. . . . I really get tired when people start calling for race. Same thing with Dominicans, they play the race card also. (Felipe, NY middle-class male, age 32)

Perhaps more injuriously, these community members become perpetrators of social practices that could act to the disadvantage of their community peers:

(11) a. Sometimes African Americans are, like, brown colored. And there are people in Dominican Republic who are brown colored. But some African Americans don't talk Spanish. So, you could tell if that person talks a lot, a lot of Spanish, you can tell that they're not African American . . . My friend looks like an African American. But he says that his mom is from Dominican [Republic], and I was, like, "Give me some words in Spanish," and he was, like, "Hola." And he says some stuff and he looks like African American, but then he showed me a picture from the Dominican [Republic] and I was, like, "Oh." (Manuel, NY middle-class male, age 11)
b. That's the only girl that's Dominican that looks African American. . . . I thought she was African American. Me and her we've been in the same class since like first grade. [And you didn't know she was Dominican?] Nope, until she started to talk Spanish, like around second grade. She starts to talk Spanish, and I'm like, "Aren't you black?" And she's like, "No, I'm Dominican. I'm 100% Dominican." I was like, "Oh, snap." I just kept quiet. (Manuel, NY middle-class male, age 13)

It should not go unremarked that the boy quoted in (11a), again in two interview sessions, requires substantiation beyond minimal language samples in accepting his black peer's self-attributed Dominican identity; his intransigence may owe to the fact that he himself would not do well in such a trial: he reports speaking little Spanish. For him, Dominican identity is not founded in language:

(12) I don't like to talk Spanish. . . . Even if I only know a little bit of words, I keep saying I'm still a Dominican Republican. . . . A Dominican who doesn't speak Spanish is still Dominican. (Manuel, NY middle-class male, age 11)

The disparate mindsets represented by this boy and the one previously quoted in (9) demonstrate that the link between language and identity is variable (Fishman 1997). For the white Dominican boy, language is marginal and optional; for the black Dominican boy, language is an essential, if not *necessary* indicator of his ethnicity (surely, in his experience, a Dominican who cannot "perform Hispanic ethnicity" through Spanish expression is African American). These data thus offer a characterization of language as both decisive and detachable; the motivation and outcome of language acts (vis-à-vis Dominican ethnic identity) depends on the social frame of the linguistic performance (cf. Bell's 1984 "audience design"). Speakers select from a repertoire of socially indexical linguistic features for identificational and relational purposes (Coupland 2001).

Resembling Without Passing

Alongside the strong affective and social factors that favor Spanish language maintenance for Dominicans in the U.S. diaspora, there exists a counteracting set of norms that attach significant importance to English, as much motivated by instrumental (e.g., economic and educational) factors as by a resistance to shouldering the blame for the "language problem" often attributed to Hispanics. Many Dominicans, especially children and young adults, will seek to distance themselves from their native language, reserving the vernacular for the intimacy and safety of the community and home; such a functional distribution of languages has become a real option for escaping linguistic prejudice and for becoming assimilated into the English-dominant U.S. society (Lippi-Green 1997).

Not surprisingly, Dominicans' posture toward English is often accompanied by a prescriptive tendency that esteems a standardized norm for the language—that of the European American majority:

(13) [Translation] I like to hear Anglos. The Anglo has a good accent and speaks clearly, clearly. The pronunciation and the vocabulary too. It's not discrimination. (Nelda, NY middle-class female, age 42)

Of course, it is a social evaluation that confers prestige on certain linguistic features and stigma on others. The woman quoted in (13) privileges the European American accent, believing that Dominicans' African appearance, especially when bolstered by African American speech characteristics, will elicit unfavorable stereotyped reactions. In fact, she discounts all varieties of English that depart markedly from the sanctioned standard. Not even Spanish–English bilingual teachers are exempted from her negative assessment; she would not enroll her children in bilingual education programs because, in her judgment, the teachers did not speak English properly. It was important, therefore, that her children speak a "good" (i.e., non–African American) English. In (14), she points unsympathetically to her stepdaughter, who was counseled by school officials to enroll in remedial speech classes where she was taught to suppress undesirable African American speech traits:[16]

(14) [Translation] The counselor asked me, "Is she Afro-American?" They gave her some speech training, and she improved a lot, and now they don't mistake her. (NY middle-class female, age 42)

As witnessed, through language performance, Dominicans are able to design and display "a network of identities" devised on the basis of the preferences and predispositions of the larger context (Coupland 2001; Tabouret-Keller 1997). However, in articulating multiple identities and goals, Dominicans may do more than simply validate societal order—they may subvert categorization. In the running account of who "we" and "they" are, the Dominican elder in (15) lays claim to attributes associated with two groups: neither "black" nor "white," she is both at once (cf. Eastman and Stein 1993).

(15) Context: Mara (age 70+) comments on Social Security benefits

Mara: Yo tengo que practicar mis palabritas en inglés para cuando yo vaya a la oficina del seguro.
"I have to practice my few words of English for when I go to the Social Security office"
Interv: ¡Como cuando usted se hizo ciudadana! ¿Y qué usted les va decir?
"Like when you became a citizen! And what are you going to say to them?"
Mara: Yo les voy a decir, "I am American."
"I'm going to tell them, [. . .]"
Context: later that same day
Mara: (To two Anglo-American passersby) Buenos días.
"Good day."

Interv: ¡Ellos no hablan español!
"They don't speak Spanish!"
Mara: Yo les dije así para que no fueran a creer que yo soy de esa gente
    negra de aquí.
"I said that to them so that they wouldn't think that I am one of those
    people from here."
Interv: (Referencing Social Security benefits previously discussed) Aquí
    se les paga igual a todos, blancos y negros también.
"Here they pay the same to all, whites and blacks too."
Mara: Yo prefiero que no me paguen.
"I'd prefer they not pay me."

As shown, Mara's language choice not only responds to her social situation, but *defines* it (cf. Giles and Powesland 1975). Nevertheless, it would be disingenuous to suggest that this elder is drawing on identity as an artifice to pass as a member of different groups from one instant to the next (cf. Coupland 2001). In truth, for this Dominican, there is no ownership of "Americanness" (and all that the term implies, e.g., patriotic and therefore English-speaking) or of "whiteness." This matter constitutes the kernel of a second important criticism against Le Page and Tabouret-Keller's "acts" proposal leveled by Coupland (chapter 1 this volume), namely, "how the acts of identity perspective interprets 'resembling,' when 'resembling and passing as' is a radically different process of social identification from 'resembling without passing.' " (In the same way, the girl referenced in (14) may activate a non–African American social identity, but may not pass for non-black.)

Moreover, the linguistic and social behavior that is represented in the extract in (15) patently undermines the opportunism inherent in the "acts" paradigm, already signaled. For though this exchange may be interpreted as reflecting the speaker's agency, for example, in projecting divergent social identities that are attuned to relationships between language forms and stereotyped social roles, it may likewise be viewed through a darker lens in which social parameters prevail— notice the symbolic gesture of relinquishing her coveted "American" status and entitlements in eschewing racial categorization: Before all else, she is non-black. Still, one might surmise that it is perhaps irrelevant whether the Spanish-language performance "plays out" for its intended audience; ultimately this language act may be self-directed. The speaker may not be projecting an identity for others, as much as reaffirming, through co-construction, a non-black identity for herself.[17] The ensuing discussion looks to other ways of "being Dominican" in the diaspora.

DOMINICAN, WITH OR WITHOUT
THE SPANISH LANGUAGE

The previous testimonials of the Dominican experience indicate that the salience of the Spanish language in the development of an ethnolinguistic identity is not a static phenomenon; rather, it depends on the context in which identity is expressed. Dominican identity in the U.S. diaspora is based on how Dominicans perceive themselves, how they strive to distinguish themselves from their black (Haitian and) African American neighbors, and how they are viewed by (European and) U.S. societies (Torres-Saillant 1995). The process of self-identification among the children of immigrants may be just as conflictual and complex, although they may not sense the isolation and hostility (cf. Utakis and Pita 2005b). Their identificational outcomes are less predictable than that of their parents and are formed in relation to multiple reference groups (in the Dominican Republic and the United States; in Spanish and English) and to the classifications into which they are placed by their native peers, schools, ethnic community, and larger society (Portes and Rumbaut 2001).

In resolving the identity "dilemmas" that may arise, multiple trajectories present themselves for second-generation children (cf. Gans 1992; Kasinitz et al. 2001; Portes and Rumbaut 2001; Waters 1999). They may assume an American identity, an option that is largely excluded, since for many Dominicans it is tantamount to becoming African American, as noted above and again exemplified in (16a). Alternatively, they may display a pan-ethnic self-identification based on language, as in (16b): they are Spanish because they speak Spanish (cf. Bailey 2000a, b). Or they may adopt U.S. government categorizations, as in (16c).[18]

(16) a. I consider myself Dominican American. [Why not just American?] Because I always speak Spanish. [Why not just Dominican?] Because I like I speak English and people think that I'm from here, cause they say I don't' look Hispanic. They say I look like I'm African American. . . . I don't get it. Cause they're like thinking I'm African American, but they're not thinking about anything else, cause all they're mostly thinking about is African Americans. They're not thinking of any other kind. So once they see me, they're like, "Oh, an African American." [And when you tell them you're Dominican?] Then they think I'm African Dominican American. I said that once, "I'm black." I wasn't thinking straight. They were like, "Okay." [They believe

you when you say you're black, but not when you say you're
Dominican?] Yes, that's the thing I don't get. They usually think
I'm black American, so I'm like plain American. (José, NY
middle-class male, age 10)

    b. [When people ask] I say I'm Spanish. [If I don't say anything]
people think you are black. . . . I think it's just because of my
color. (Gus, NY middle-class male, age 11)

    c. I have my friend, named Francis, she's a girl. She's twelve, in
seventh grade. She looks African American, but she's Hispanic.
(Manuel, NY middle-class male, age 13)

Yet another, more prevalent option is for second-generation U.S.
Dominicans to retain their parents' national identity, with or without
a hyphen.

(17) a. I am both: Dominican and American. 'Cause I was born in
America and my mom was born in Dominican [Republic] and
my dad too. . . . The blood you have in you is Dominican. (Gus,
NY middle-class male, age 11)

    b. I consider myself Dominican American. . . . [I can be
Dominican even if I don't speak Spanish.] I can prove it since
my parents have Dominican passports. I can bring it, like,
"How's this proof?" (José, NY middle-class male, age 10)

    c. I consider myself Dominican. I never thought of [hyphen-
ation]. (Andrés, middle-class male, age 13)

For these youths, being Dominican goes beyond the symbolic, to
include a large measure of national loyalty and ethnic pride, as
recounted in (18).

(18) a. I'm proud to be Dominican. If you're Dominican, be
Dominican. Represent your country . . . I have a Dominican
girlfriend. We met at a rally for Dominican Independence Day:
February 27. (Manuel, NY middle-class male, age 13)

    b. [Translation] My son, who must be third generation
[American], will tell you that he's Dominican. [He listens to]
perico ripiao, waves the Dominican flag and everything. It must
be because we maintain all of our culture: the food, the music,
the Spanish language. (Pedro, NY middle-class male, age 33)

Levitt too notes that since many children spend their preschool years,
summers, and even parts of their adolescence in the Dominican
Republic, "the customs, values, and traditions of these countries
become ingrained in their everyday vocabulary and can be activated or

deactivated at different stages of the life cycle" (2001, 21). For not unlike their parents, for whom return to the island is not a myth, but mandate, the children of the second generations return to the island for regular visits—for Holy Week and Christmas, summer vacations, and family weddings and funerals—likewise settling into this Dominican/American identity, "the state of mind that permits them to remain actively linked to life in the native land while also becoming acclimated to the values and norms of the receiving society" (Torres-Saillant and Hernández 1998, 156).

(19) I go every year [to the Dominican Republic]. Sometimes I think, yeah, I wish I would move back there. If you don't go there for a while, it just doesn't feel the same. You just feel different. You can't live in the U.S. your whole life and be Dominican. You have to go once, at least once a year. (Carlos, NY working-class male, age 14)

With loyalty to their Dominican heritage, the Dominican boys interviewed pledge allegiance to the Dominican vernacular—especially for communicating with family and friends abroad (cf. Dicker and Mahmoud 2001). And yet, despite the importance of the heritage language for U.S. Dominican youths, the official goal of most language-education programs in New York City is to teach immigrants English without regard for the development of the native language (cf. Pita and Utakis 2002b). There is substantive evidence that these boys' heritage language is eroding: Language decay was evident at the levels of the lexicon, the syntax, and narrative constructions. Thus, though they suggested that ethnicity and language were inextricably linked, the specific form of the language was of little concern to them. Nevertheless, the Spanish language remains a commodity to be exploited for emblematic purposes, reappropriated and renegotiated, and most meaningfully, shared with nonnative peers:

(20) a. We just talk like fast and sort of like in slang. Like if I'm with my friends, I'd be like, "Dímelo loco." . . . You can't be Dominican without speaking Spanish. (Andrés, middle-class male, age 13)
b. Dominicans, Puerto Ricans, and some African Americans like to speak Spanish; they play around with it. (Manuel, NY middle-class male, age 13)
c. Mostly it's all the Spanish kids talking together, and some of the white kids get in too. They learn a lot of Spanish, the white kids. (José, NY middle-class male, age 10)

The interviews with the youths further exposed, rather suggestively, that just as their parents' language practices appear to be fading, so too do their parents' racial attitudes.[19] For unlike first-generation immigrants who formulated their impressions of African Americans by observing them at a distance, the second generation takes in additional data with which to construct a more informed view (21):

(21) a. I went out with one black girl. It was easy, it was fine; we got along and everything. I didn't tell my parents nothing. I only tell my parents sometimes who I go out with. (Manuel, NY middle-class male, age 13)

b. I have mostly white and black friends [in the United States]. I think I could have black friends in the Dominican Republic. (Gus, NY middle-class male, age 11)

c. I wouldn't mind if they [schoolmates] thought I was black. I would just say, "No, I'm not." (Andrés, middle-class male, age 13)

It remains to be determined whether this demeanor represents an awakening to their racial consciousness. It is noteworthy, however, that these gestures do not include an embrace of African American social identities or language patterns, contrary to the findings reported in Bailey (2000a, b).[20]

In summary, rather than seeking expressions of status and prestige, as defined in the U.S. setting, the majority of the adolescents interviewed appear to manifest a solidarity with their black and white peers and a new discourse of intimacy with their compatriots. For these youths, "being" Dominican in the diaspora extends beyond the application of self-label for self-categorization to the communication of a new, more inclusive Dominican narrative.[21] In doing so, they advance toward dismantling essentialist concepts of Dominican identity (as non-black, Spanish-speaking, etc.)

CONCLUDING COMMENTS

To summarize, the behaviors exemplified and the theories in which these have been couched have disclosed linguistic acts of ethnicity among U.S. Dominicans and the social structures in and through which identities are formed and performed. The testimonials have additionally illustrated that through the indexicality of their languages, U.S. Dominicans can simultaneously confirm and contest the identities foregrounded in the wider sociocultural frame as well as project new identities in (re)constructing sociocultural contexts.

In addition to affording insights into the linguistic dimensions of Dominican society, this enterprise may most profitably be understood as a point of departure for future research. Indeed, as the present study represents a sampling of local case studies, any statement of the significance of the findings cannot be but somewhat guarded; further examination of the role of language in the ongoing process of the construction of ethnic identity is certainly warranted.[22] For example, researchers may want to determine how identity is negotiated in diasporic settlements such as Puerto Rico, in which the Spanish language is not contrastive. These observations additionally call for added attention to the linguistic and social attachments portrayed by first- versus second-generation immigrants on the basis of their experiences with linguistic and racial discrimination in the diaspora. The investigation likewise invites a comparison of the findings from individual interviews with those yielded by group interviews; these may further illuminate the issue of overt and covert prestige assigned to language practices among adults versus children and male versus female family members. The study also enjoins researchers to examine the indexicality of specific Spanish language forms (e.g., dialectal versus standard) and English language forms (e.g., African American vernacular, Dominican English, standard American English) in distinct settings, for example, urban versus suburban, enclave versus other. Finally, the work provokes further consideration of the attitudes and ascriptions of the larger community toward Dominican cohorts.

NOTES

* I would like to express my gratitude to Janina Brutt-Griffler and Catherine Davies for their invitation to participate in the English & Ethnicity Symposium and in the resultant book project. I would also like to thank Nikolas Coupland for having shared the unpublished manuscript that greatly informs this work and John Rickford for his encouraging comments on the larger undertaking.

1. I adhere to Fishman (1997, 329), in using the term "ethnicity" to denote group "belongingness."

2. Also dedicated to the examination of the linguistic dimensions of society is *macro-sociolinguistics*, which studies what societies do with their languages, e.g., the attitudes and attachments that account for the function and distribution of speech forms in society, language shift, maintenance, and replacement, and the delimitation and interaction of speech communities (cf. Coulmas 1997).

3. Fundamental in communication science is the distinction between three communicative objectives: instrumental, interpersonal, and identity. The latter objective, involving "the management of the communicative situation to the end of presenting a desired self-image for the speaker and creating or maintaining a particular sense of self for the other (Clark and Delia 1979, 200, cited in Rickford 2002)" is theoretically and analytically central to the present work.

4. Of course, there are other potential contributory factors to language performance—audience, topic, etc.—even when identity is held constant (Rickford 2002).

5. Approximately one million Dominicans live in the United States, concentrated largely along the eastern seaboard, with an estimated 500,000 residing in the New York metropolitan area (2000 Census). I refrain from using the term *Dominican American*, as many self-identify as *Dominican*, irrespective of citizenship or place of birth. A different perspective is offered in Pita and Utakis (2002), who distinguish a hyphenated identity, associated with a past in one country and a present and future in another, from a bi/transnational identity, which spans two countries.

6. Since the 1990s, the discourse of Dominican studies has come to be characterized by two competing projects: one oriented toward community building among Dominican immigrants and the other locating its focus on transnational practices among members of this group.

7. Consult Otheguy et al. (2000), Valdés (2000), and Zentella (2000) for thoroughgoing overviews on the language situations of Cuban Americans, Mexican Americans, and Puerto Ricans, respectively.

8. Dominican Spanish differs from other varieties of Spanish with respect to lexical, phonological, morphological, and syntactic features (cf. Toribio 2000a,b, 2001).

9. Zentella (1997) found that Dominicans expressed highly negative opinions about Dominican Spanish, and 80 percent said that Dominican Spanish should not be taught in schools.

10. The response in (2b) is especially telling, as the speaker promotes the "Spanish" dialect without definite knowledge of its Iberian origin.

11. In addition to the weakening and deletion of syllable-final /s/ and /n/, of note are the outcomes of phonological processes that affect syllable-final liquids: /l/ and /r/ in checked position may be rendered as [l] in the capital city of Santo Domingo, as the off-glide [i] in the northwestern agricultural countryside of the Cibao Valley, and as [l] in the southern region. Of these forms, the lateral liquid of the *capitaleño* carries the greatest cultural and political capital, as expected, and the northwestern *cibaeño* pronunciation the least.

12. These are commonly buttressed by reference to related desirable or undesirable physical characteristics, such as hair textures and size of the nose, lips, hips, and buttocks.

13. One significant result of the racial discrimination that Dominican immigrants experience in the United States is the fortification of social

investments and cultural constructs that ensures their full participation in both home and host societies. Sontag and Dugger (1998) report that Dominicans are perhaps the most transnational of all New York immigrants. The articulation and reinforcing of transnational practices enables Dominicans to circumvent racial barriers and be recognized "for who they truly are" (Levitt 2001; Pita and Utakis 2002).

14. Levitt reports that almost twice as many Dominicans in Massachusetts classified themselves as "other" than as "black" on the 1990 Census; and more than 80% of all Dominicans classified themselves as "white" between 1996 and 1999.

15. It is unquestionable that such moments of social insecurity heighten children's consciousness of what it means to be black and Dominican in the United States, and contribute in large measure to language loyalty (Toribio 2003).

16. Of course, though the advantages of adopting normative speech patterns may be obvious to educators, the matter is far from transparent for many children, and all that may be effected by teachers' exhortations in the direction of uniform standards is an increase in any linguistic insecurity that pupils already sense (cf. Lippi-Green 1997; Wardhaugh 1998).

17. This is conformance with Coupland's (2001) exhortation toward a shift in focus in sociolinguistic studies from addressee responsiveness to "identity management" and "self evaluation" (cf. Giddens 1991).

18. Graham notes that the strong residential concentration of Dominicans in New York has lessened the need to form coalitions with other ethnic groups; Dominicans thus have been able to focus on a national, not a broader, ethnic identity (cf. Duany 1994).

19. There is considerable evidence that the notion of race will become more refined as immigrants and their descendants attempt to create and manage their new identities.

20. These differential findings may owe to the contexts of investigation: Bailey examined Dominican identity in a Dominican enclave in Providence; the children referenced in this text are members of suburban communities in New York.

21. Unlike their parents who forged a reactive ethnicity in the face of racial discrimination, these youths demonstrate a mode of ethnic identity formation that is not adversarial.

22. Many of these issues will be pursued in a large-scale cross-disciplinary study entitled, "The (Re)Construction of Identities: Dominicans in New U.S. Destinations," elaborated in collaboration with four colleagues: Jeffrey Cohen (Anthropology), Gordon de Jong (Sociology, Leif Jensen (Rural Sociology), and Salvador Oropesa (Sociology).

REFERENCES

Bakhtin, M. 1981. *The Dialogic Imagination*. Austin, TX: Austin University Press.

Bailey, B. 2000a. Language and negotiation of ethnic/racial identity among Dominican Americans. *Language in Society* 29: 555–582.

Bailey, B. 2000b. Language and ethnic/racial identities of Dominican American high school students in Providence, Rhode Island. Unpublished doctoral dissertation, University of California, Los Angeles.

Baud, M. 1997. "Constitutionally white": The forging of a national identity in the Dominican Republic. In *Ethnicity in the Caribbean: Essays in Honor of Harry Hoetink*, G. Oostindie (ed.). London: Macmillan Caribbean. 121–151.

Bell, A. 1984. Language style as audience design. *Language in Society* 12: 145–204.

Cambeira, A. 1997. *Quisqueya la Bella: The Dominican Republic in Historical and Cultural Perspective*. Armonk, NY: M.E. Sharpe.

Clark, R.A. and J. Delia. 1979. *Topoi* and rhetorical competence. *The Quarterly Journal of Speech* 65: 187–206.

Coulmas, F. (ed.). 1997. *The Handbook of Sociolinguistics*. Cambridge, MA: Blackwell.

Coupland, N. 2001. Language, situation, and the relational self: Theorizing dialect-style in sociolinguistics. In *Style and Sociolinguistic Variation*, P. Eckert and J. Rickford (eds.). Cambridge, UK: Cambridge University Press. 185–210.

Dicker, S.J. and H. Mahmoud. 2001. Survey of a bilingual community: Dominicans in Washington Heights. Paper presented at the 23rd Annual NYS-TESOL Applied Linguistics Conference.

Duany, J. 1994. *Quisqueya on the Hudson: The transnational identity of Dominicans in Washington Heights*. Dominican Research Monographs. New York: CUNY Dominican Studies Institute.

Eastman, C. and R. Stein. (1993). Language display: Authenticating claims to social identity. *Journal of Multilingual and Multicultural Development* 14: 187–202.

Eckert, P. 2000. *Linguistic Variation as Social Practice*. Oxford: Blackwell Publishers.

Fishman, J.A. 1997. Language and ethnicity: The view from within. In *The Handbook of Sociolinguistics*, F. Coulmas (ed.). Cambridge, MA: Blackwell. 327–343.

Gans, H. 1992. Second-generation decline: Scenarios for the economic and ethnic futures of the post-1965 American immigrants. *Ethnic and Racial Studies* 15: 173–193.

Gal, S. 1988. The political economy of code choice. In *Codeswitching: Anthropological and Sociolinguistic Perspectives*, M. Heller. New York; Mouton de Gruyter. 245–264.

Gal, S. 1995. Language and the "arts of resistance." *Cultural Anthropology* 10: 407–424.

García, O., I. Evangelista, M. Martínez, Carmen Disla, P. Bonifacio. 1988. Spanish language use and attitudes: A study of two New York City communities. *Language in Society* 17: 475–511.

Giddens, A. 1991. *Modernity and self-identity: Self and Society in the Late Modern Age.* Cambridge: Polity Press.

Giles, H. 1977. *Language, Ethnicity and Intragroup Relations.* London: Academic Press.

Giles, H. 2002. Couplandia and beyond. In *Style and Sociolinguistic Variation*, P. Eckert and J. Rickford (eds.). Cambridge, UK: Cambridge University Press. 211–219.

Giles, H. J. Coupland, and N. Coupland. 1991. *Contexts of Accommodation.* England: Cambridge University Press.

Goffman, E. 1974. *Frame Analysis.* Harmondsworth: Penguin.

Graham, P. 1998. The politics of incorporation: Dominicans in New York City. *Latino Studies Journal* 9: 39–64.

Graham, P. 2001. Political incorporation and re-incorporation: Simultaneity in the Dominican migrant experience. In *Migration, Transnationalization, and Race in a Changing New York*, H. Cordero-Guzman, R., Smith, and R. Grosfoguel (eds.). Philadelphia: Temple University Press. 87–108.

Grasmuck, S. and P. Pessar. 1996. Dominicans in the United States: First- and second-generation settlement, 1960–1990. In *Origins and Destinies: Immigration, Race, and Ethnicity in America*, S. Pedraza and R.G. Rumbaut (eds.). Belmont, CA: Wadsworth. 280–292.

Gumperz, J. 1972. *Directions in Sociolinguistics: The Ethnography of Communication.* New York: Holt, Rinehart and Winston.

Gumperz, J. 1982. *Discourse Strategies.* Cambridge, UK: Cambridge University Press.

Irvine, J. 1989. When talk isn't cheap: Language and political economy. *American Ethnologist* 16: 248–267.

Itzigsohn, J., C. Dore Cabral, E. Hernández medina, and O. Vásquez. 1999. Mapping Dominican transnationalism: Narrow and broad transnational practices. *Ethnic & Racial Studies* 22: 316–340.

Kasinitz, P., J. Battle, and I. Miyares. 2001. Fade to black? The children of West Indian immigrants in Southern Florida. In *Ethnicities: Children of Immigrants in America*, R. Rumbaut and A. Portes (eds.). New York: Russell Sage Foundation. 267–300.

Labov, W. 1972. *Sociolinguistic Patterns.* Philadelphia: University of Pennsylvania Press.

Landale, N. and R.S. Oropesa. 2001. White, black, or Puerto Rican? Racial self-identification among mainland and island Puerto Ricans. *Social Forces* 8.

Le Page, R.B. and A. Tabouret-Keller. 1985. *Acts of Identity: Creole-Based Approaches to Ethnicity and Language.* Cambridge: Cambridge University Press.

Levitt, P. 2001. *The Transnational Villagers.* University of California Press.

Lippi-Green, R. 1997. *English with an Accent: Language, Ideology, and Discrimination in the United States.* New York: Rouledge.

Macaulay, R. 1975. Negative prestige, linguistic insecurity and linguistic self-hatred. *Lingua* 36: 147–161.

Macaulay, R. 1991. *Locating Dialect in Discourse*. New York: Oxford University Press.

Milroy, L. 1980. *Language and Social Networks*, 1st ed. New York: Basil Blackwell.

Milroy, L. 1987. *Observing and Analyzing Natural Language*. Oxford: Basil Blackwell.

Moya Pons, F. 1981. Dominican national identity and return migration. Occasional Papers #1. Gainsville, FL: University of Florida Center for Latin American Studies.

Otheguy, R., O. García, and Roca, A. 2000. Speaking in Cuban: The language of Cuban Americans. In *New immigrants in the United States*, S.L. McKay and S-L.C. Wong (eds.). Cambridge, UK: Cambridge University Press. 165–188.

Pita, M.D and S. Utakis. 2002. Educational policy for the transnational Dominican community. *Journal of Language, Identity and Education* 1, 14: 317–328.

Portes, A. and R. Rumbaut. 2001. *Legacies: The Story of the Immigrant Second-Generation*. New York: Russell Sage Foundation.

Rampton, B. 1995. *Crossing*. London: Longman.

Rickford, J. 2002. Style and stylizing from the perspective of a non-autonomous sociolinguistics. In *Style and Sociolinguistic Variation*, P. Eckert and J. Rickford (eds.). Cambridge, UK: Cambridge University Press. 220–231.

Rodríguez, C. 2000. *Changing Race: Latinos, the Census, and the History of Ethnicity in the United States*. New York University Press.

Royce, A. 1982. *Ethnic identity: Strategy of diversity*. Bloomington, IN: Indiana University Press.

Sontag, D. and C.W. Dugger. 1998, July 19. The new immigrant tide: A shuttle between two worlds. *The New York Times*, A1, A28–A30.

Sorensen, N.N. 1997. There are no Indians in the Dominican Republic. In *Siting Culture: The Shifting Anthropological Object*, K. Fog Olwig and K. Hastrup (eds.). New York: Routledge. 292–310.

Tabouret-Keller, A. 1997. Language and identity. In *Handbook of Sociolinguistics*, F. Coulmas (ed.). Cambridge, MA: Blackwell. 315–326.

Toribio, A.J. 2000a. Minimalist ideas on parametric variation. In *NELS 30: Proceedings of the North East Linguistics Society*, M. Hirotani, A. Coetzle, N. Hall, and J.-Y. Kim (eds.). Amherst, MA: University of Massachusetts. 627–638.

Toribio, A.J. 2000b. Setting parametric limits on dialectal variation in Spanish. *Lingua* 110, 315–341.

Toribio, A.J. 2001. Language variation and the linguistic enactment of identity among Dominicans. *Linguistics: An Interdisciplinary Journal of the Language Sciences* 38, 1133–1159.

Toribio, A.J. 2003. The social significance of language loyalty among black and white Dominicans in New York. *The Bilingual Review* 27, 1: 3–11.

Torres-Saillant, S. 1995. The Dominican Republic. In *No longer Invisible: Afro-Latin Americans Today*, Minority Rights Group (ed.). London: Minority Rights Group.

Torres-Saillant, S. 2000. *Diasporic Disquisitions: Dominicanists, Transnationalism, and the Community*. Dominican Studies Working Paper Series 1. CUNY Dominican Studies Institute.

Torres-Saillant, S. and R. Hernández. 1998. *The Dominican Americans*. Westport, CT: Greenwood Press.

Trudgill, P. 1974. *The Social Differentiation of English in Norwich*. Cambridge: Cambridge University Press.

Trudgill, P. 1983. *On Dialect: Social and Geographic Factors*. Oxford: Basil Blackwell.

Trudgill, P. 1986. *Dialects in Contact*. New York: Basil Blackwell.

Utakis, S. and M.D. Pita, 2005. An educational policy for negotiating transnationalism: The Dominican Community in New York City. In *Reclaiming the Local in Language Policy and Practice*, A.S. Canagarajah (ed.) Mahway, NJ: Lawrence Erlbaum Associates.

Valdés, G. 2000. Bilingualism and language use among Mexican Americans. In *New immigrants in the United States*, S.L. McKay and S-L.C. Wong (eds.). Cambridge, UK: Cambridge University Press. 99–136.

Wardhaugh, R. 1998. *An Introduction to Sociolinguistics*. Malden, MA: Blackwell.

Waters, M. 1999. *Black Identities: West Indian Immigrant Dreams and American Realities*. Harvard University Press and Russell Sage Foundation.

Woolard, K. 1985. Language variation and cultural hegemony: Toward an integration of sociolinguistics and social theory. *American Ethnologist* 12, 738–748.

Zentella, A.C. 1997. Spanish in New York. In *The Multilingual Apple: Languages in New York City*, O. García and J. Fishman (eds.). Berlin: Mouton de Gruyter. 167–201.

Zentella, A.C. 2000. Puerto Ricans in the United States: Confronting the linguistic repercussions of colonialism. In *New Immigrants in the United States*, S.L. McKay and S-L.C. Wong (eds.). Cambridge, UK: Cambridge University Press. 137–164.

Zentella, A.C. 2002. Spanish in New York. In *The Multilingual Apple: Languages in New York City*, 2nd edition, O. Garcia and J. Fishman (eds.). Berlin: Mouton de Gruyter. 167–201.

PART 3

CONTEXTS

6

Speaking for Ourselves: Indigenous Cultural Integrity and Continuance

Simon J. Ortiz

"*Chuunah ehnu, chuwah-stah-nih-eeh, gaimeh-eh.*" *Ahmoo stih-naaya*, beloved my mother, would ask us to go for water at the *chuunah*. The little river that carries water from the Zuni Mountains flows through lava beds to our lands in the village area of *Deetseyaamah*. *Deetseyaamah* means north door because of the road that goes between two mesas north of the mother Pueblo of *Aacqu*. *Deetseyaamah* was where I was raised. Our gardens and fields and orchards needed the water brought by the *chuunah*. We needed the water for our home use. And *shrah-dyaiyuutyai-meeshee*, our animals, like the plow horses Charley and Bill and the goats and chickens, needed water, the *tsih-tsee* that was brought to us by the *chuunah*.

My brothers and sisters and I walked down the dirt trail to the *chuunah* and brought it home in pails. We poured the water into fifty-gallon, metal drum barrels until they were full so we could use the water for washing dishes, washing clothes, household cleaning, and such things. By the time I was a boy, the *tsih-tsee* from the *chuunah* was already badly contaminated from the upstream small town of Grants that dumped its raw sewage into the *chuunah*; so nobody drank the water flowing in it. And soon the uranium mining and processing industry would begin in the 1950s that would absolutely pollute and poison the *chuunah* so that it was hardly usable as a river any longer. But we still irrigated our orchards, gardens, and fields with it.

"*Nah-chamah, nuu shtuuweetaah-stih*," I used to hear my father say, announcing he would be irrigating our fields and gardens with the *tsih-tsee* coming from the *chuunah*. The *chuunah* was the only source of surface water, except for the rainwaters flowing off the nearby mesas, hills, and upper plateau areas when it thunderstormed. We loved to go with him when he went to "water" the chili, carrots, cilantro, lettuce, radishes, beets, whatever was in our garden that season. I loved the blue, yellow, and white flowers growing along the narrow banks of the waterways leading to our garden. The flowers were exquisite and I was always happy to see them. My brothers and sisters and I would find bluish-purple and pink flowers that were shaped like little ducks. When you plucked the flowers from their stems and put them in the water they floated downstream just like ducks too! We giggled as we watched them bob along on the fast moving water.

In the spring before every planting season, the *tsih-tsee mayah-rrdomo* or *hoochanee* would call men and boys of the Acoma people to come to the main irrigation ditches to clean them of debris, trash, old weeds, and repair breaks in the *kuupaashtuweetsah* so they would be fully prepared for the upcoming planting and growing season's use. All of the *Aacqumeh hanoh*, even ones who didn't always farm gardens and fields, were expected to help. That was the way it was: you were expected to help because you were part of the *hanoh* and the *hanoh* was you. *Ahyaamaatse*, that was the rule. To help, that was the rule. *Uumaatse utrahnih*. To be of help, that was the way to be. When I think about it, the Acoma language is a confirmation of this concept. The *dzahtyawah hoochani* would come among the people—years back in olden times they'd stand on the highest point in the community to say what they had to say—and announce the day you were to be at the starting point of the ditch work. And you better be there!

To be of help—*uumaatse utrahnih*. Ultimately, because of the vital and essential role that indigenous language has in indigenous culture and community that is significant. Without that role the Acoma community would not be what it is. The *Aacqumeh hano* were expected—or were required!—to help with the maintenance of the *kuupaashtuweetsah* so the lifewater would continue be provided by the *chuunah*; likewise the Acoma language used by the *dzahtyawah hoochani* in his announcement was an example and confirmation of *uumaatse* or help that was needed.

It is now many years later. I was born in 1941. World War II had already started in Europe when I was born but it would not be until December of that year that the United States would enter it. In 1941, most of the livelihood at *Deetseyaama* and *Aacqu* was subsistence farming: corn, beans, pumpkins, melons, vegetables, apple and peach orchards, alfalfa fields. There was plenty of water running in the *chuunah* that was diverted into the *kuupaashtuweetsah*, which were located above garden and field areas. On the way to both our gardens and fields, we crossed the *chuunah* where a railway trestle spanned it. Sometimes right there, just at the moment we were crossing on the railroad tie bridge across the *chuunah*, a passenger troop train would be passing. The heavy steel wheels and diesel engines of the AT&SFRY train would shake the railway trestle and the ground below our feet. Even the water in the *chuunah* would tremble in tiny little waves.

I would wave and wave my arms and hands at the *son-dah-rrotitrah* in their uniforms as they headed west for California and beyond. When I remember that time I always remember the *chuunah*. But it is more than memory that is apparent. It is obvious to me that remembrance has limitations. It is not enough to have a memory of the *chuunah*; memory is unreliable. Although you cannot live your life all over again and you cannot undo traumatic change, you can still expect that the *kuupaashtuweetsah* you construct will carry water from the *chuunah* to your gardens, fields, and orchards. Your work in constructing the *kuupaashtuweetsah* is absolutely important and significant. Its importance and significance is in making use of your knowledge so that the *kuupaashtuweetsah* does not become mere memory.

Dzah dzeenah hamanah skah-ow-dimah tyanuh. You hear that more often now. We don't plant now anymore. It's said with longing I believe, and nostalgia. Today there are fields still along the *chuunah* but they are scattered now in patchwork patterns. There used to be broad expansive, contiguous fields of tall corn and lush alfalfa. Acres and acres. Especially fields that *Stieu-rrlu* and his sons had on the north side of both the *chuunah* and the railroad tracks. And also Salvador and his sons who planted, raised, and harvested wheat. When I was a nine-year-old altar boy I would see on the walls of the local Catholic church sacristy photos of cornfields in the 1930s and 1940s: Acoma men standing with their horses and wagons alongside their fields. *Dzah dzeenah.* No more today. *Hamanah.* No longer now. *Skah-ow-dimah-tyianuh.* Planting we. Although I used the terms "longing" and "nostalgia" a moment ago, I don't mean them in a sentimental way; rather I mean for them to be statements of factual circumstance and condition more than anything else. *Yuunah kaatya shra hanoh tyaimishee, amoo babaahtitra eh nanatitra*—in time past our people, beloved grandmothers and grandfathers—did plant and cultivate corn, pumpkins, beans, and other food crops. They sowed the sandy soil of the lands immediately around *Aacqu*, the mother pueblo, twelve miles south of *Deetseyaamah*.

When I was a boy I remember my father, mother, a couple of my aunties, and their families—children, lots of children always, among whom I numbered—helping my grandpa in his bean fields east of the tall stone cliffs of *Acqu*. Hot June or July days. Blue endless sky. We had canvas water bags hanging from the wagon sides that we kept visiting every few minutes. Joyful work, family, land, community. *Haatse* and *hanoh*. Land and people. *Haatse eh hanoh*. How could we not feel it was absolutely important we share our lives with each other? It was necessary to feel that *shrau-yuugaiyih-shee* was what kept the land, rain, plants, animals, children, *stsai-dzee* all tied together. *Stsai-dzee.* Everything. *Stsai-dzee.* We don't plant like we used to anymore. Nostalgia and longing in the voice? Yes. However, you don't like to admit that's what the feelings are: nostalgia and longing. But we have to admit to them.

Kqow-kuwah sruweh meh-yuunah wai hamaa-dzeh-shi-meh skuwah dhawaah? How and in what way still presently today do we make use of past knowledge? That's a very important question. It has to do with a cultural consciousness that continues today in the way we are conscious of language use in ordinary everyday activities and experiences. And this applies to activities and experiences within the immediate indigenous cultural community context of *Deetseyaamah* and *Aacqu* as well as other locales, that is, wherever *Aacqumeh hanoh* are living today.

Some years ago I was in Cullowhee, North Carolina, doing a poetry reading at a local university. During the reading I noticed a young indigenous woman with a pre-teenage boy sitting in the front row of the crowd. Afterward as I signed books, they approached me and the woman said, "*Guwaadze*," holding out her hand. "*Dawaa-eh*," I said as I shook her hand. And then I asked, "*Kqow how yeh-emih kuudah?*" She told who she was and added, "Alice Rose *stsee-naya eh* Grandma Juana *stah-dya-ow.*" We were related to each other! What was an Acoma woman doing way over there in North Carolina? What was an Acoma poet doing way over there reading poetry? It turns out she was married to a man from the Cullowhee region, which is the homeland of the *Tsalagi*, known usually as the Cherokee who are the original owners of lands in what are now the states of Georgia, North Carolina, South Carolina, Tennessee, even Florida, and perhaps even Alabama. "*Kehstee-dzuu-chuutah?*" I asked her, wondering if she went home to Acoma once in a while. "*Hah-ah. Drutyuh?*" she said.

Kee-haamah, kqow ku dze 1950 or 1951, ai Ku-ku-pana nu stowdimah. Grandma Juana lived near there, her house and her daughters' houses at the foot of the sandstone cliffs-coves called

Ku-ku-pana. Back then, perhaps it was in 1950 or 1951, we planted at Owl's Cove. Alice Rose was just a girl then, kind of skinny, probably no more than eight or nine years old. I remember Sandoval her father, a man who was a World War II veteran I idolized. He always wore a leather bomber flight jacket from the war, and he worked at the small rural airfield *ai-deeniyaa-tsakuuwah*—at a place on the mesa above us to the south. The *kuupaashtuweestah* that diverted the *chuunah* water at Anzac miles to the west ran between the cornfield we planted and the several houses where Grandma Juana and her daughters, their husbands, and their children lived.

We make use of past cultural, community, and clan knowledge because that knowledge informs us of our present-day lives. *Amoo-uh Baabah* Juana passed on to the spirit world a number of years ago where she will continue to live on forever. And Sandoval, her in-law, also passed on. In the early 1950s he was killed in a car wreck on a curvy road running through the volcanic lava beds near *Ku-ku-paanah.* I shall always remember Sandoval who worked at the little airfield—*kee tuu-duutruh-shra koowah-guhnaatah ka* leather flight jacket. I have to say that past knowledge is more than just past knowledge; it is knowledge that impacts upon and determines our present reality.

Aie-shrah-trutyiae shtee-dyanuu ai-ehme chuunah dzah. From our house down northward, that's where the *chuunah* is. It flows from the west, from *Hee-shah-mi Kuuti* in the Zuni Mountains through the Bluewater Valley and through the town of Grants. "*Maameh gkai-kah aneh tseechuu dawah kahnaatrutaiyah*," my mother used to say, telling us "Very fortunately, it was beautiful, big, and good as it ran along." I imagined the *chuunah* in the time of my mother's childhood: it is a vigorous, clear-running stream of water flowing through Grants eastward, eventually meandering through the volcanic lava beds on the western border of the Acoma Pueblo reservation. However, by the time I was a student at Grants High School, the uranium mining and processing industry was getting started. In less than a decade during the late 1950s, the *chuunah* had become terribly polluted from the raw sewage that Grants and its growing population were dumping into it. It was being poisoned by uranium industry tailings and harsh chemical wastes flowing by the tons into the water table that supplied the *chuunah.* Vast depletion of *chuunah* waters was being caused by the processing plants and the underground uranium mining.

"*Haa-dih shrutahnitrah?*" was a common question we asked each other during that time. Where do you work? Meaning: who do you work for? "*Kcqee* Kermac *stihtaanitrahni.*" Or Anaconda Corporation. Or Homestake-Sapin Partners. Or Phillips Petroleum. We, indigenous peoples of Acoma and Laguna Pueblos and the Navajos of Prewitt and Bluewater, were low-income or no-income poor people, simply cheap labor, who didn't seem to have much choice. No longer self-sufficient subsistence farmers, numbers of us went to work in the uranium industry. We were laborers for the most part or lower-echelon skilled workers, never anything in management. Even as we may have complained about the uranium industry affecting the quality of our water and causing the depletion of it, making the *chuunah* waters unfit for human consumption, we didn't quit working for the uranium industry. We had no choice. "*Hah-tsuumah aie-shruhtanitrah?*" How long have you worked there? "*Mai-kqu wa ka-aitrah.*" A long time, it's been.

Recently I heard *Stah-naweh* Maurus say, "I remember the day the *chuunah* died. I was down at the *chuunah* the day before and it was okay then. The next day when I saw it, I couldn't believe it! But I had to. I really couldn't believe it. But it was true. It was dead." And he described how the river had been turning gray and brackish and sluggish. Wildlife such as muskrats and ducks and other birds had diminished over recent years. "And it was smelling sour, not like the river anymore, not like years ago," he said. There used to be good trout in it. You'd see them in shallow stretches of the *chunnah*, especially where it narrowed on slippery hard clay bottoms and where it was shallow and clear. Big rainbows. Beautiful, glittery bodies. There used to be large, stately cottonwoods and tall bushy trees we called *perritoh*, which had sort of sweet pulpy, gray berries. Russian olives, I think. Years before, when *stah qkuie-trah yaanih dzeshi* Linda *eh* Russell had not yet been married long, we'd have picnics on weekends at a favorite fishing spot. Sometimes along with the trout we'd hook fat round-bellied suckers that usually smelled "too fishy" although we'd fry them up and eat them anyway. The *chuunah* was a good place to be. We all thought that and even said it with such remarks such as, "It's really nice here, ainnit" or "*Maameh aneh-tsah wai*" as we sat in the shade of a grove of tall slender *kahnimaah* eating our picnic lunch of baloney and tortillas and oatmeal cookies.

When I recall my nephew speaking about the day the *chuunah* died, I'm almost overcome by longing and nostalgia. *Keegai-kah hamah wehmeh aneh dzah*, I think. *Keegai-kah hamah wehmeh aneh dzah.* I almost don't want to repeat in English what I just said in the *Aacqumeh kah-dzeh-nih-neeya.* But I will. Back when it used to be good that way. *Keegai-kah hamah wehmeh aneh dzah.*

Simply because I grew up with the *chuunah*, I cannot conceive of an *Aacqumeh* world without the *chuunah*. It was a crucial and essential component of the landscape of my childhood. And it remains a component of the landscape even though I am no longer a child and I don't live at *Aacqu* fulltime. The *chuunah* and its waters is the lifeblood of plants, animals, the land, the people, and the world I've known since I was a child. And the *chuunah* is actually more than that; it is a crucial component of the cultural life experience of the agrarian culture the *Aacqumeh hanoh* know. And even more than that since it is central to the all-encompassing energy my community of *Aaqumeh hanohtitra* knows. Literally, the ecological and cultural landscape I thrived within growing up at *Aacqu* would not be Existent without the *chuunah*. Without the *chuunah* I knew as a child and know now at the present time, I would not have a concept of *Aacqu* as the cultural and geographical place that is my home.

Like the *chuunah* is a crucial component of the cultural and physical life of the *Aacqumeh hanohtitra*, the *Aacqumeh tsehnih* is critical to the Existence of the cultural community of *Aacqu*. When I think of culture and language, I have to say this: culture and language do not thrive unless these crucial elements are in place side-by-side and complementary to each other. Language is vital to culture; there is no cultural Existence because the community cannot thrive without it. There is a problem therefore when the prominence of the English language is exercised predominantly. It is more often than not the only community language used; all or most communication from family conversations to tribal governmental business is conducted in the English language, not indigenous languages. English has pushed indigenous languages out of the indigenous family, culture, and community, and this has brought about inevitable conflicts that run the gamut of intra-family relationships, tribal governance, and education. This problem and conflict has resulted in damming the flow of cultural and community continuity. Without this flow, the indigenous culture and community is stifled and affected negatively, much like the physical and cultural life of the *Aacqumeh hanohtitra* of *Deetseyaamah* was diminished when the water of the *chuunah* was polluted and depleted by the uranium mining and processing in the 1950s through the 1980s.

Tuu-nee. Knowledge. *Uu-tuu-nee.* What one knows. One's knowledge. *Shrow-tuu-nee.* What we know. Our knowledge. We know the world by what is around us, that is, by what we are within. And we know the world by how we relate to what is around us. The *chuunah* flowing with its waters through *Deetseyaamah* brings life to us, the

Aacqumehtitrah, so that we may thrive. As long as the *chuunah* flows, we thrive. Knowledge of ourselves is crucially involved with what is around us, how we relate to it, and how we participate with it. When *amooh steenayah* asked me to bring buckets of water from the *chuunah* for our household use and I did, I was involved with, relating to, and participating in the cultural and geographical world of *Aacqu.* When my father irrigated our gardens and fields with *tsih-tsee* from the *kuupashtuweestah,* which came from the *chuunah,* he was involved with, relating to, and participating in the cosmos of the Acoma way of life.

Recently I was emotionally struck, even astounded, when I heard my eldest sister Linda say to one of her grandchildren, *"Amoo dya-ow,* you must not waste that water, *amoo-uh tsih-tsee,"* referring to the water being beloved just like she, her granddaughter, was beloved. The vocal tone of her mixed English-Acoma comment pertained both to her precious beloved granddaughter and the water she was being told about. This same tone of voice and message can be applied when talking about plants, animals, landscape features, weather conditions, and so on because of their inclusion within the cultural philosophy of the Acoma community. That means the regard for the *chuunah* and the *tsih-tsee* it provides is very significant and must not be carelessly treated. When the *Aacqumeh* cultural community bases its very Existence upon such regard and reverence, there is no question that the very material and concrete existence of the *chuunah* and its waters is necessary.

If anything this is the point of my consideration and conclusion: indigenous cultures and communities such as the *Aacqumehtitrah* are going to continue integrally intact when the resource of their languages is not polluted, diminished, stifled, and prevented from useful and free expression.

7

ENGLISH AND THE CONSTRUCTION
OF ABORIGINAL IDENTITIES IN THE
EASTERN CANADIAN ARCTIC

Donna Patrick

INTRODUCTION

One of the major preoccupations of Aboriginal educators and of a growing number of linguists is the plight of indigenous languages internationally. Given the consequences of colonization and globalization, many of the world's minority languages appear to be threatened by the increasing use of English, French, Spanish, or other dominant languages within particular nations (Crystal 2000; Dorian 1989; Fishman 2001; Grenoble and Whaley 1998; Krauss 1992; Nettle and Romaine 2000). In Canada, where fifty or so aboriginal languages are still spoken, only four (Cree, Ojibwe, Dakota, and Inuktitut) are said to be "truly viable" (Kinkade 1991). A more optimistic estimate places this figure at seven, due largely to the language revitalization efforts that took hold in the 1990s. Whatever the actual number, Aboriginal language use in Canada has been rapidly decreasing, and many local community leaders and educators have become involved in language revitalization and maintenance initiatives specific to the circumstances of their communities.

In this context, a great deal of ethnolinguistic research on indigenous languages has focused on the indigenous languages themselves and on the links between language, ethnicity, and identity. Such a focus has often meant that the complementary investigation of the dominant language—which addresses its interaction with Aboriginal languages as well as its relationship to "ethnicity" and "identity" in Aboriginal contexts—has been neglected. This has arisen largely because the minority (and often threatened) indigenous languages have been seen to warrant the greater attention.[1] The research to be reported here, however, can be seen as a modest attempt to redress this imbalance, in its investigation of both indigenous and dominant language use in three northern Canadian communities: one in Nunavik, the Inuit region of northern Quebec, and two in Nunavut, the Inuit territory created in 1999 out of the eastern part of the Northwest Territories. In the northern Quebec community investigated, two indigenous languages—Cree, an Algonquian language, and Inuktitut, an Eskimo-Aleut language—are spoken alongside French and English. In the Nunavut communities, English is gaining dominance over Inuktitut in several contexts and especially among younger speakers. In this chapter, I discuss the meanings and everyday use of English and Inuktitut in the Eastern Canadian Arctic and, to a lesser extent, of French in northern Quebec. My focus, though, will be English and the paradoxical role that it plays in these contexts: having the status of a language of colonization and dominance, yet at the same time serving as a necessary tool for the assertion of Inuit land rights and autonomy and for the protection of Aboriginal languages, rights, and local institutional control within the Canadian state.

In the eastern Canadian Arctic, use of Inuktitut has remained robust, displaying a vitality not found in Inuit, Eskimo (i.e., Alaskan Inuit), or Yup'ik communities further west. The strength of Inuktitut in the eastern Arctic can be understood only in a broader historical, cultural, political, and economic context, and with respect to local sociolinguistic constraints on language choice in everyday communication (Patrick 2001, 2003a, b). In northern Quebec communities, for example, the relative isolation of the Inuit subsistence economy and its rather late integration into the trade and wage-labor market economies have been factors in the continued use of Inuktitut in everyday life. When settlement and schooling became mandatory—in the 1950s and 1960s, respectively—there was still very little English spoken in community life. In the 1960s, with the politicization of Aboriginal groups in Canada and internationally, the processes of modernization and decolonization in the eastern Arctic gained momentum. The promotion of Aboriginal rights, particularly through spokespeople educated in federal English-language schools, became the focus of the Aboriginal leadership. Thus, in a relatively short time, modernization—both material, involving the sedentarization of Inuit and their transition from a nomadic lifestyle to one based in government-created settlements, and political, involving the pursuit of land claims—radically altered the face of these northern communities.

In the multilingual reality of northern Quebec and Nunavut, English and, to a lesser extent, French are important resources, not only in the construction of local identities and ethnicities, but in the pursuit of greater autonomy for these regions. The goal of achieving autonomy has involved the negotiation of land claims settlements, the struggle for control over local institutions and for a greater say in economic development projects, and the assertion of the right of Aboriginal peoples (or First Nations, as these groups have come to be known in Canada) to traditional hunting and other harvesting activities. This goal involves not only achieving greater local control over community and regional affairs, institutions, and economic development, but also improving the social conditions of Aboriginal citizens and gaining the means to ensure the social, cultural, and linguistic "survival" of the group.

In the pursuit of increased institutional, political, and economic control, Aboriginal groups have turned their attention to education and the role of English, French, and Aboriginal languages in schools. What has become clear to these groups is that self-government, and the increased autonomy that goes with it, requires expertise among their political leaders and in the administrative and other professional positions in local communities. This has led to a fundamental paradox for many Aboriginal groups. On the one hand, indigenous peoples have recognized the need to mobilize politically in order to be able to maintain their ethnic "difference" as reflected in their distinctive cultural, economic, and linguistic practices. On the other hand, these groups have needed to engage in thoroughly modern political processes, involving the use of English or other dominant languages, in order to secure land rights and obtain the financial resources needed to maintain this "difference"—even though acquiring the ability to engage in such processes threatens to undermine the very "difference" that justified mobilization in the first place.

In what follows, I describe the consequences of this paradox and the meanings that English holds for Inuktitut speakers in eastern Canada. To begin, I discuss the notions of ethnicity and identity and how these categories are socially constructed through processes of language choice and discursive practices. Next, I turn to an examination of Canadian colonial history and the role of English in northern Quebec and Nunavut, going from nineteenth-century contact with English-speaking traders and missionaries to twentieth-century political developments. I then examine contemporary multilingual language practices in northern Quebec and bilingual practices in Nunavut, as documented through ethnographic research in these regions. This will give a good idea of the various roles that English plays in this region, which include helping to construct contemporary Aboriginal identities and, for some, helping to resolve tensions between economic development and material well-being, on the one hand, and centuries-old indigenous cultural and linguistic practices, on the other.

ETHNICITY AND IDENTITY

In discussing "ethnicity" in relation to the construction of the "black subject," Stuart Hall remarks:

> If the black subject and black experience are not stabilized by Nature or by some other essential guarantee, then it must be the case that they are constructed historically, culturally, politically—and the concept which refers to this is "ethnicity." (Hall 1992, 257)

For the purposes of this discussion, I assume that ethnicities are constructed on the basis of shared histories, traditions, and language; and that ethnicity, as a social construct, is an important element in shaping personal and cultural identities. Since the notions of "ethnicity" and "cultural identity" are distinct but still closely related, it is worth emphasizing here that the former is a social construct that relates a given individual to a particular ethnic or racialized group or groups through their engagement in particular cultural practices and beliefs; whereas the latter pertains to the way that individuals identify themselves on the basis of their cultural background and practices including religion, social class or caste, linguistic background, country of origin or residence, and other aspects of cultural identification. In multicultural and multilingual settings, ethnic and cultural identities are in constant flux: individuals position themselves or are positioned by others, and their membership in particular groups is socially constituted through interaction. The boundaries between social groups are thus fluid and not "fixed"; and ethnicity, as a social category, is "achieved," not prescribed or biologically determined (Hensel 1996, 84).

The notion of "achieved" or historically and socially constructed ethnicity is in opposition to the notion of "ethnic absolutism"—an essentialist view of culture that views ethnicity in terms of homogeneous, static "ethnic essences" that define who one is and shape one's character (Gilroy 1987; Rampton 1995). Ethnicity, on the latter view, is taken to have "an exclusive emphasis which hides all the other social categories which individuals belong to," such as age, gender, class, and residence. Ethnic absolutism thus obscures the often complex and contradictory relationships of solidarity and conflict "across a *range* of category memberships" (Rampton 1995, 8).

Hall's (1992) notion of "new ethnicities" recognizes a diversity of cultural and symbolic practices "which work with and through difference," and are thus "able to build those forms of solidarity and identification which make common struggle and resistance possible" and to do so "without suppressing the real heterogeneity of interests and identities" (Hall 1992, 254–255, cited in Rampton 1995, 287). As such, "ethnicity" is based on a constellation of shared practices and of interaction among members of social networks, which together form the boundaries of ethnic groups.

Members of ethnic groups share views on cultural identity and cultural difference in relation to other groups. At the same time, the group can sustain a good deal of diversity. According to Barth (1969), an ethnic group inhabits an "ecological niche" where the group and its boundaries are constituted in material, social, and political conditions. Beliefs, attitudes, and symbolic and material cultural practices—including, but not limited to, those related to language, kinship, religion, and dress—are taken up in different ways and shaped by the processes of boundary formation and maintenance.

Regarding Yup'ik Aboriginal identities in southwestern Alaska, Hensel (1996) notes that individuals are ethnically identified not by what language they use or by physical attributes, but by social practice. In particular, ethnicity and cultural identity are constructed in terms of how one engages in subsistence activities—hunting, fishing, preparing meat, and the like—and, in a community where most residents speak more English than Yup'ik, how one uses English to "talk about" the meaningfulness of these practices. This discursive construction in turn constructs "Yup'ikness" or Aboriginal ethnicity and local forms of cultural identity.

In contemporary Canada, bureaucratic and legal discourses have been instrumental in constructing the broad social categories of "nativeness" and "whiteness" (or "nonnativeness"). These categories have been constructed historically, through colonial discourses about the "Other" and through political economic processes that have brought Aboriginal peoples into a world economic system. The processes in question were associated first with the fur trade and then with sedentarization—that is, forced settlement and assimilation—which in Canada was effected through residential schooling, agricultural training, and the bureaucratization that resulted from the implementation of the Indian Act in the late nineteenth century.

Since the 1970s, the politicized term "First Nations" has operated across ethnic, linguistic, and historical boundaries, uniting Native peoples in a struggle with governments and multinational corporations over land rights in Canada and elsewhere. However, although the unifying and hegemonic categories of native and nonnative or "Aboriginal" and "non-Aboriginal" construct and consolidate two distinct ethnic and cultural realities, they can also be seen to "overlap." This is evident in the distinctive Metis (or "mixed") identities that have arisen from the fur trade unions between European traders and Aboriginal women (Dickason 2003) and the contemporary linguistic practices among Aboriginal peoples on southern Canadian reserves and in urban environments (Darnell 2004). How Euro-North American practices have been taken up by indigenous peoples and vice versa—and the impact that this has had on language practices—are questions that warrant further investigation. The following sections can be seen as contributing to such an investigation.

COLONIAL AND MORE RECENT HISTORY

Hundreds of years of contact, domination, and resistance have produced vibrant and complex personal and cultural identities in Aboriginal communities in Canada, in which English and French as well as indigenous languages have come to play a large part. Given this history, a good understanding of the current symbolic, sociopolitical, and economic roles of English, French, and Inuktitut in the eastern Arctic requires some knowledge of the region's colonial history. Considerations of space preclude detailed discussion here (but see, e.g., Patrick 1994, 2003b), hence the following sketch should suffice for our purposes.

A good place to begin is with the establishment in 1670 of the Hudson's Bay Company in London, England, which led to the creation of a number of trading posts in northern Canada throughout the eighteenth and nineteenth centuries. This, in turn, led to a significant economic shift for Aboriginal peoples in the region: away from a subsistence economy and toward a trading economy. Moreover, since most traders were from the British Isles, this also led to the introduction of English into the region.

It is worth noting that this shift was not without friction. In particular, the establishment of the more northerly trading posts met with Inuit resistance; and many of these posts ended up closing and reopening again during the course of the Hudson's Bay Company's efforts to pursue trade with the Inuit. Trading posts had already been established along the Hudson's Bay coast in the eighteenth century, yet Inuit did not enter into trade, to exchange caribou, seal, and fox skins, until around 1840 (see Francis and Morantz 1983 for a detailed history of the Hudson's Bay Company in the region). We can take this date, then, to mark the time at which the Inuit hunting economy in the Hudson's Bay region, as in other areas, began to be co-opted into the world capitalist system through the fur trade (Wolf 1982).

Missionaries followed traders into the region soon after trading posts were set up. Like the traders, most missionaries spoke English as a first language. Unlike the traders, however, they made fluency in Inuktitut a priority, in order to be able to carry out their work more effectively. The missionaries who ventured into the Canadian North believed that the "Natives" could access the "word of God" more directly in their own language, and so took it upon themselves to teach Aboriginal peoples to read and write in their own language, so that they would be able to read the Bible and prayers. (This ideology also governed missionary work in other colonial settings; see Meeuwis 2000.) In Arctic Quebec, missionaries from the Church of England spread the use of a syllabic writing system, based on shorthand, which was originally developed in the mid-eighteenth century by James Evans, an English missionary working with the Cree. The Cree system was adapted for Inuktitut by the missionary John Horden in the mid-nineteenth century and adapted further by Rev. Edmund Peck, who arrived in the Hudson's Bay area in 1876 and who was already familiar with the Inuktitut (Eskimo) grammar books compiled by Moravian missionaries on the Labrador coast (Harper 1983; Marsh 1964). The Inuit syllabary that was created consists of forty-five symbols: three vowels and forty-two consonant–vowel combinations.

Literacy in syllabics spread quite quickly, because it was fairly easy for one person living on the land to learn the syllabary and to pass on this knowledge to others. By the late nineteenth century, Peck had moved to an area in the Baffin region (close to Iqaluit, now the capital of Nunavut), and brought both Christianity and literacy in the syllabic system to the Inuit living there. With these literacy practices inevitably came certain values of moral regulation, including the notion that writing is a "civilizing" force, necessary for development and modernization (Street 1984, 183–188), and is associated with greater intelligence and cognitive skills (Olson 1986; Olson et al. 1985). Also noteworthy here is that the syllabic system, introduced by Europeans and tied to the value that they placed on reading and writing, has ironically become a significant symbolic resource for expressing Inuit identity in the eastern Arctic (see Shearwood 2001). Moreover, this system has, over the past century, also become an important element in the construction of Inuit political, educational, and bureaucratic structures.

A second shift in the local political economy of Inuit living along the Hudson's Bay and other Arctic coastal regions came with forced sedentarization, English-language schooling, and the introduction of wage labor in the late 1950s and early 1960s. Inuit entered the cash economy by helping to construct and maintain community infrastructure and, in some cases, military or other government buildings. In Great Whale River, a settlement on the eastern Hudson's Bay coast, an army base was constructed in 1955 as part of the American-sponsored DEW Line, the Distant Early Warning radar system, which spanned the Canadian Arctic. This and other military installations built across the Arctic brought with them an increasing number of English speakers.

For over three centuries, then, traders, whalers, missionaries, and government, law enforcement, and military personnel have brought with them European linguistic and cultural practices such as writing and formal schooling as well as material goods such as tea, flour, sugar, cloth, rifles, and ammunition, which have quickly become integral to Inuit society. As regards the languages that were brought to the region, it was English that was the dominant language during most of this three-century period, associated as it was with those in charge of the posts and with the church, the government, and wage labor. However, French became increasingly prevalent in northern Quebec during the 1960s—a period that saw a rise in Quebec interests in the region and in the role of French as a language of the state, business, and administration. Not surprisingly, the presence of French in this region has had a significant effect on the lives of Inuit there, with French figuring alongside Inuktitut and English in education, government publications, and services.

The 1960s are worth highlighting in our brief review of the history of the Inuit in this region, since it can be seen to mark the beginning of a third shift in Inuit history, characterized by increased political mobilization. The greater French interest in northern Quebec and the rise of Quebec nationalism not only brought more French-speaking government agents to the region, but also led to a consciousness-raising among Inuit regarding their rights, especially their language rights. In the early 1960s, Inuit won the right to be schooled in Inuktitut, if only for the first one or two years of school (Patrick and Shearwood 1999). Inuit also sought access to Inuktitut-language radio programming in their communities and to Inuktitut-language media in general. In the 1970s, further mobilization around issues of land rights emerged with the Quebec government's interest in hydro-electric development on Cree and Inuit territory. The court cases that arose eventually led to the ratification in November 1975 of a landmark land claims settlement, the James Bay and Northern Quebec Agreement (JBNQA), touted as the first "modern" land claims settlement in Canada.

The signing of the JBNQA has not been the only significant result of political negotiation for Inuit. Another has been the creation of Nunavut in April 1999 out of the eastern portion of the former Northwest Territories. This feat required thirteen years of negotiations, from 1976 until 1989, for the drafting of an agreement-in-principle on the land claim itself, and another three years for ratification of the final agreement (Creery 1994, 141). Nunavut, which is a territory approximately one-fifth the size of Canada, thus represents not only a land rights settlement but also a new territory (Kusugak 2000, 20). This is not the case for Nunavik, where an agreement for greater political and economic autonomy within the province of Quebec has yet to be finalized (Nunavik Commission 2001).

Land claims and other political achievements won through negotiation have thus been a preoccupation of First Nations groups in Canada for more than three decades—as has a preoccupation with the connection between language, ethnicity, and political action. It is no accident that these two preoccupations emerged at about the same time. In the 1970s, as negotiations for the James Bay Agreement and Nunavut began and as Euro-Canadian and Aboriginal relations were becoming more openly politicized, relations between Quebec and Canada were also becoming more politicized. This was a period when language, ethnicity, and politics began to figure prominently in Canadian politics. The federal government had passed the Official Languages Act in 1969, making Canada officially bilingual in English and French—a move to accommodate French-speaking Quebec within a predominately English-speaking Canada. That same year, a preliminary statement on Indian policy, known as the White Paper on Indian Policy, was tabled in Parliament (Burnaby 2002, 78). This statement proposed to abolish "Indian status" in Canada and, in effect, to abrogate the rights and obligations of the federal government toward Aboriginal groups as outlined in treaties and in the Indian Act, first enacted in 1869 and revised in 1876. This policy statement overlooked the perspectives of the Aboriginal leaders themselves, who were quick to condemn it. This, in turn, gave impetus to Aboriginal politicization and set the stage for a new era of land claims negotiations, as Aboriginal Canadians sought to gain more control over their land and welfare, highlighting the linguistic and cultural "distinctness" that motivated their desire for such autonomy. The result of the JBNQA and the Nunavut land claim has been the establishment of Inuit government and economic organizations, the common goal of which is to give Inuit more control over their territory. This includes a say in resource management and development and potential profit-sharing in economic ventures on their land.[2] The creation of the political entities of Nunavik and Nunavut has also resulted in language policies and practices that legitimize English, French, and Inuktitut (and a fourth Aboriginal language, Inuinnaqtun, spoken in the western part of Nunavut). These languages have been granted regional "official" status in education, government publications, and services and have become significant parts of the linguistic, cultural, economic, and political landscape.[3]

These results suggest that political mobilization, despite its obvious successes, has also introduced further complications into the lives of the Inuit living there, which I describe in the next section.

LAND CLAIMS, LANGUAGE, AND SELF-GOVERNMENT

The complications just alluded to lead inexorably to the conclusion that the political mobilization initiated by First Nations groups in the early 1970s has, in fact, created a fundamental paradox for them, which has also been recognized in Aboriginal communities elsewhere. English has been the language needed for engaging in modern political processes, which have involved legal and media campaigns to secure land rights and to obtain the financial resources necessary to promote and maintain local indigenous language and culture. However, the high value of English—and other dominant languages in other national settings—has, in certain cases, actually undermined the status and viability of local languages and cultures, which political mobilization had sought to protect in the first place. One case of this, as documented by Bunte and Franklin (2001), is that of the Paiute of Arizona, among whom English usage has been increasing, whereas Paiute usage has been decreasing. Ironically, this increase in English has coincided with the politicization of the Paiute in the 1980s and 1990s, the goals of which were to gain recognition for their tribe from the Bureau of Indian Affairs and to settle a land claim. These struggles have played a large role in the modernization of these communities and have resulted in better health care, education, and other social programs. Yet the central role of English in the realization of these goals has placed Paiute in a more vulnerable position.

In the eastern Canadian Arctic, the effect of political mobilization has been arguably less damaging to Native languages, but certainly no less significant. The powerful role of English in politics and in the economy has had a great impact among the Inuit of both northern Quebec and Nunavut. For example, the English used by Inuit leaders, acquired largely as a result of the residential schools in which they and other native children received their primary and/or secondary education, has turned out to be of real benefit in the political processes that led up to the signing of the JBNQA and the Nunavut agreement. The hardships that such schools imposed on students, their families, and their communities—separating children from their families and communities, and permitting them to speak only English, thus hindering their first-language development as well as their acquisition of "traditional" knowledge and practices—should not be underestimated. That said, both the mastery of English and the forging of friendships among Native students in these schools emerged as critical for political mobilization.

A key difference, however, between the role of English in the political mobilization and modernization of the Inuit and its role in the political mobilization and modernization of such groups as the Paiute, as described above, is that in the former case, increased English usage has not resulted in reduced Inuktitut usage among Inuit community members. That is, despite its status as a "modern" Western language, English does not appear to be displacing Inuktitut, the "traditional" language of the Inuit. Although there are sociohistorical, political, and economic explanations for the presence of rapid language shift in some communities and not in others, one factor that stands out among the Inuit is their broad acceptance of the need to "modernize" their language so that it can cope with new terminologies, technologies, and concepts of modern government and social change.

Inuit political and institutional leaders have been instrumental in promoting a "modernized" Inuktitut to help them in realizing their political and economic goals and in fulfilling the mandate of their positions. This modern Inuktitut has been developed by "language workers" engaged in a new language economy. These workers, who have included translators, interpreters, educators, and journalists, have quickly developed new, standardized Inuktitut vocabulary that can readily express the modern concepts associated with new technologies and with contemporary political, legal, and social institutions. As it happens, this vocabulary, and the concepts that it accesses, is needed not merely by an elite but by Inuit society more generally, once more facing profound social changes. A clear demonstration of this need being felt by the community as a whole was the requirement that everyone have access to the contents of land claims agreements and institutional proposals, in the form of documents and other communications in Inuktitut, in order to be able to support or reject particular proposals, as they were expected to do. The roles of both Inuktitut and English here highlight the fact that this and other Aboriginal contexts have demanded that Aboriginal leaders not only display fluency in a dominant language, but also have legitimacy as respected and "authentic" (i.e., locally accepted) members of Aboriginal communities. This dual linguistic and cultural requirement, together with the prestige associated with leadership and other powerful positions in modern Aboriginal economies, has given a high symbolic value to Inuktitut as well as English skills, reflecting the paradoxical situation that Inuit find themselves in today. An appreciation of this paradox, already mentioned above, seems basic to an understanding of contemporary Inuit society in Canada. Since such an appreciation is most easily gained by considering current language practices in this region, we turn to this in the next section.

LANGUAGE PRACTICES IN NUNAVIK:
ENGLISH, FRENCH, AND INUKTITUT
IN NORTHERN QUEBEC

In Inuit communities in Nunavik, three language groups, representing English, French, and Inuktitut speakers, vie for power, each seeking to win or retain legitimate control over a territory that currently has about 9,000 Inuit and a few hundred non-Inuit residents. English-language speakers, both Inuit and non-Inuit, tend to justify the dominant role of English in the region in terms of its historical significance and its national and international role in communication—including communication with other Aboriginal peoples. English is, of course, the language of globalization and of international markets, and holds a hegemonic cultural, political, and economic position. It is also one of the official languages of the federal state; and Inuit, as Canadian citizens, have found that knowledge of English is necessary for dealing with federal government offices regarding such matters as taxes and pension benefits. The historical role of English and its current role nationally and internationally thus insure its dominance in the territory. But the dominance of English is only one small part of the picture of language use in Nunavik communities.

We can get a better idea of how certain patterns of language use emerge in these communities by taking a closer look at one such community, that of Great Whale River, which was the site of ethnographic research reported in Patrick (2003b). This community, situated on the border between Nunavik and the Cree territory to the south, has about 600 Inuit, 600 Cree, and 100 nonnative residents. Because of its history and geographical location, English, French, Inuktitut, and Cree are all spoken there on a daily basis. What this means when bilingual or multilingual speakers meet is that they must make a choice about the language of communication. This choice can be described in terms of certain basic tendencies. One such tendency is for two people who speak the same first language to speak in that language, even if they share a second or third language. Accordingly, two Inuktitut speakers will tend to speak Inuktitut and two Francophones French, even if each also speaks English. This pattern of language use cuts across settings and participants, and across network types, whether related to work, friendship, or family. It also persists despite the historical dominance of English and the use of this language in intercultural communication. Another tendency is for a conversation to begin in either French or English when two non-Inuit interlocutors meet and the language background is not known. In these cases, French is often the likelier choice given that most nonnatives in northern Quebec are Francophone (i.e., of French-speaking background) rather than Anglophone (i.e., of English-speaking background).

As it happens, English is spoken to some degree by almost everyone in the community, and functions as a lingua franca in intercultural communication—that is, between speakers of different languages. The wide distribution of English is, as noted earlier, a legacy of the English-language schools set up for Inuit and Cree by the federal government in 1958. It is also the second language of the educated Aboriginal elite, who have pursued their studies in southern Canadian high schools, colleges, and universities; and of Francophone managers and government employees, who need English for their work in the North. In this community, English is also the language used in settings where people from all four language groups congregate. These settings include the gymnasium, the hockey arena, and the airport, where signs, schedules, and the like are written in English.

Although English has the highest number of speakers in the community, fewer than 1.5 percent of residents, and fewer than

15 percent of the nonnative population, speak English as their first language. Whereas some residents are more fluent than others, everyone beyond a small number of Inuit and Cree elders uses some form of English in intercultural encounters. These encounters take place all over the community—at the gymnasium, the arena, the store, the school, the post office, and the health clinic. In other words, the use of English has become integral to community life.

This can be seen from the interview data that were part of the investigation of English, French, and Inuktitut among Inuit youth and bilingual adults reported in Patrick (2003a). Among such data were responses such as the following one to the question of how Inuit were coping with the growing tension between French, English, and Inuktitut in their lives.

> I think English is used rather too much because we are in a situation where we are in Quebec, and French is the language of the majority of the province, and being Inuk, trying to maintain our culture and our language, and we're having to deal with three languages, everybody tends to turn to English. (Inuk Municipal Council employee, quoted in Patrick 2003a, 129)

> Being fortunate enough to speak two languages, I can communicate in either one. . . . When I communicate with the Cree, for example, or the French, the communication is in English, and that's no problem, everybody seems to speak English. (Inuk Municipal Council employee, quoted in Patrick 2003a, 132)

In these and other interactions with Inuit, a clear message emerged that English was a highly valued symbolic resource, which permitted communication with others locally, nationally, and globally. Mastering the language also improved employment opportunities, since knowledge of English was required for any workplace that dealt with the public.

People I worked with, socialized with, or interviewed also offered certain less obvious reasons for including English in the community fabric and for assigning it an important function in the construction and maintenance of groups and social identities. One such reason is that English has become an attractive language for many young people, not only for its association with North American popular music, television, film, and the Internet, but also for its role in peer group interaction and the construction of emerging youth identities. The importance of this factor was highlighted by an interaction (recounted in Patrick 2003b, 184–185) that I had while teaching adult education in a village on the Ungava coast. This was with a student in her early twenties, who now wanted desperately to learn English. The reason why, she told me, was that two of her classmates, friends she had grown up with in another village, would use English words and phrases when they were together, significantly limiting her ability to participate in their conversation and thus excluding her from it. From this perspective, knowledge of English can be seen as a status symbol, representing not only a form of knowledge acquired in school and a language of the external, "Southern," culture to which Inuit youth have been increasingly turning, but also a means of "belonging" to and identifying with a group. In other words, English has become a key force in the construction of emerging Inuit youth identities and peer groups. It is this close connection between English and "Southern" culture—and the media and technologies that access this culture—that seems to be behind the increased use of English and has thus been of great concern to language activists wishing to preserve Inuktitut and other indigenous languages.

Learning and speaking English can also be linked to identity in other ways. As already mentioned, English is an integral part of Canada's colonial history, having for centuries held a dominant position as the language of traders, missionaries, and bureaucrats, and of various cultural and economic practices that Inuit themselves engaged in. Indeed, speaking English as a second language was even identified with being Inuk, according to one woman I interviewed:

> Since I'm Inuk, English is my second language. The English came here and they wanted us to learn how to speak English. I really want to understand very much English. I have nothing against French; if I ever need a translator my son will translate for me. (Patrick 2003b, 133)

This excerpt suggests, then, that the long-standing presence of English in northern Quebec has led in the emergence of new ethnolinguistic identities in this region.

Of course, the dominance of English or French also has more practical consequences for Inuit—in particular, as regards their employment opportunities. Given the linguistic complexities of northern Quebec, where the second language of most Inuit remains English but where French is the official language of the provincial bureaucracy (and the language spoken by the vast majority of southern Quebec residents), knowledge of both languages is necessary in at least some workplaces and highly desirable in others. This is suggested in the following interview excerpts:

> *D:* Do you need French to do your job?
>
> *J:* Yeah, we are in Quebec, all the papers are in French . . . everything is in French, [but] for northern Quebec we translate everything in English.
>
> *D:* Do you do that translation?
>
> *J:* No, no no. I do my own paper in English, and the court proceeds in English . . . We do that because there is a principle, an agreement saying that for northern Quebec, it is recognized in the James Bay Agreement, that everything has to be translated in English. (Nonnative court worker, quoted in Patrick 2003b, 144).

> The reason why I decided to learn French is because I want to get a better job, more higher, like let's say where I'm working right now as a cargo loader, I need French in order to be working at the counter with the customers, the public. . . . I guess in some areas we need French, especially where I'm working, to deal with the public, you need English and French, and we do need the Cree and the Inuktitut, they are all important. (Cree woman married to an Inuit man, living in the Inuit community, quoted in Patrick 2003b, 145.)

English, as these interviewees point out, remains the favored second language of Inuit, although French has assumed an increasingly important position in the community, particularly in the workplace. At this point, however, knowledge of French among Inuit remains limited, leading to stopgap measures such as the translation of French documents into English to cope with this change.

For many Inuit, then, English and Inuktitut continue to be sufficient for most communicative purposes in the community. More importantly, it is these two languages that are tied to local identities, with English, to some degree at least, also being tied to Inuit youth culture. What this means is that "mixed" or hybrid bilingual identities are emerging among Inuit; such identities are contingent and emerge from the multiple shifting relationships among Inuit, Cree, French, and English speakers in this community.

Given the role that language plays in the construction of these and other Inuit identities, understanding the factors involved in language choice becomes central to an understanding of identity formation in general. What is revealed by research on cross-cultural interactions in different community settings is how language choice can be used strategically to negotiate power relations or to create solidarity, and how language can construct social boundaries that serve to include or exclude members of particular ethnic groups. For example, in northern Quebec, more traditional activities such as the community fishing derby or Christmas celebrations involved very little use of English, which resulted in the exclusion from such activities of those who did not speak Inuktitut. But there are also inclusive language practices in other community contexts.

One incident recorded in my field notes occurred in the airport waiting room, where a French-speaking Inuk was addressed in French but responded in English as a means of including in the conversation non-French speakers who were also present.

The function of language choice to include or exclude potential participants in a conversation can also be seen in the constraints that I observed on the use of French in particular social settings. One clear demonstration of such constraints involved a French-speaking teacher addressing a French-speaking Inuk student in French while the two were playing badminton in the community's gymnasium. The teacher offered the following greeting: "*Salut Anna. Comment ça va?*" (Hi Anna. How are you?) The response of the student, who was standing with two of her friends on the court, was simply "*Bien*" (Well). After this, she immediately switched back to Inuktitut to talk to her friends. From the teacher's perspective, Anna had missed an opportunity to use her French outside of the classroom. Yet, we can guess that she switched to Inuktitut in order to avoid excluding her friends, which continuing in French would have done, since they did not speak French. We can also guess that a lengthier interaction would likely have ensued if Anna had been addressed in English, the lingua franca of the community. In other words, the cost of forming an allegiance with a Francophone teacher, in French, seemed to be too high if this was to be at the expense of solidarity with her friends (Patrick 2001, 306).

As the research reported here has suggested, language used in daily contact between Inuit and nonnative residents of northern Quebec can either create boundaries—for example, during harvesting activities and community functions, and in intimate groups of friends—or remove boundaries—for example, in situations where communication between members of different groups is desirable and, in particular, where English is chosen as the lingua franca. In all of these cases, language choice is linked to issues of power and solidarity, and can be better understood only through detailed sociolinguistic investigation, of which the interview and observational data reported here can be seen as first steps.

LANGUAGE PRACTICES AND
IDENTITY IN NUNAVUT

In the previous section, we examined the situation of Inuktitut in the Inuit region of northern Quebec known as Nunavik. In this section, we turn to the region of Nunavut ("our land"), a vast area with an Inuit population of approximately 23,000, representing about 85 percent of its total population. One reason for turning our attention to this region is that the situation of its Inuktitut speakers and their language is intriguingly different from that of their counterparts in Nunavik, despite substantial similarities between the two. In particular, although the overwhelming majority of inhabitants in both regions are Inuit who continue to speak Inuktitut, the rate of Inuktitut–English bilingualism in Nunavut is substantially greater than it is in Nunavik. Another reason is that the linguistic situation in Nunavut, unlike that in Nunavik, has already been the subject of long-term investigation—in particular, by Dorais and his colleagues (e.g., Dorais and Sammons 2000, 2002), who have documented Inuktitut and English use in two Nunavut communities, Iqaluit and Igloolik. The findings of this research, though they must be approached with great caution beyond the context of Nunavut itself, can nevertheless give us some idea of the trends that already exist or are likely to exist in Nunavik. Moreover, the cautious optimism that has been expressed by residents and researchers alike regarding the continued vitality of Inuktitut in Nunavut, given continued institutional support, offers some reason for similar optimism about the future of Inuktitut in Nunavik.

In their recent study, Dorais and Sammons investigated English and Inuktitut language use in two Nunavut communities: Iqaluit, the capital and largest settlement, and Igloolik, a smaller, more northerly, and more traditional community. Despite substantial differences between these communities—for example, 60–65 percent of the 6,000 Iqaluit residents are Inuit, compared to about 95 percent of Igloolik's 1,200 residents—the vast majority of the Inuit population of each continues to speak Inuktitut (see, e.g., Dorais and Sammons 2000, 93 for details). At the same time, the rate of Inuktitut/English bilingualism in each community is high, particularly among those under the age of fifty.

In an important sense, then, these two languages are basic to the fabric of everyday life in both Nunavut communities. This conclusion emerges from various findings of the Dorais and Sammons study. These include the finding that the use of English increases with age, as children receive more English instruction in school—Inuktitut being, as in Nunavik, the medium of instruction for only the first three years of schooling—and as they speak more English with their siblings and peers. Given their fluency in Inuktitut, acquired in the home and in the first years of their schooling, bilingualism thus becomes the norm for older children and young adults. Some see clear roles for Inuktitut as well as English. As one interviewee put it, "English is a world language, hence its usefulness, but when we are among Inuit, Inuktitut is more useful" (Dorais and Sammons 2000, 106). Others favor English more generally; the results of interviews with those under the age of thirty educated predominantly in English revealed some who felt that it was easier to express their feelings and inner thoughts in English (Dorais and Sammons 2000, 107; 2002, 90).

Other findings of Dorais and Sammons's study confirm this bilingual picture, indicating that English figures prominently in the workplace especially, whereas Inuktitut plays a greater role in Inuit social networks, which construct and maintain Inuit identity. Investigation of eight workplaces in the public and private sectors revealed a high degree of bilingualism, with English generally used more frequently than Inuktitut (Dorais and Sammons 2000, 105). This is consistent with the results of interviews with Nunavut residents, which revealed that a large number of those under thirty felt that English had become more useful than Inuktitut, due largely to its ability to help residents "find a good job and support your family" Inuktitut (Dorais and Sammons 2000, 106). These results highlight the increasing use of English and its higher value in the labor market of Nunavut.

At the same time, Dorais and Sammons found Inuktitut still playing a pivotal role in the region as an aspect of Inuit identity, and starting to play a greater role in the labor market. Not only did many residents over thirty make a connection between Inuktitut and Inuit identity in interviews, but even more revealingly, Inuktitut was found to be the language that parents in both Iqaluit and Igloolik used most often or exclusively when talking to younger children, even when only one parent was Inuk. Moreover, many of the thirty to fifty year olds interviewed felt that Inuktitut would become more prevalent in the workplace given the establishment of Nunavut and the promotion of Inuktitut in government affairs (Dorais and Sammons 2000, 106). This belief may figure in parents' continued transmission of Inuktitut to their children notwithstanding the widespread use of English among older children and adults—although this use of Inuktitut may have more to do with the value that it still holds for many Inuit as a language of intimacy and of "being Inuit" (Dorais and Sammons 2000, 108).

Despite a certain overlapping of domains, then, which is observable in the use of Inuktitut and English in Nunavut, the latter is still largely the language of practical matters, and the former is still the language imbued with local meaningfulness, which serves to express Aboriginal identity. This indicates that the link between language and identity, and between Inuktitut use and the promotion of cultural values, remains strong in this region. It also indicates that opposing cultural and economic forces still create an environment favoring the use of both Inuktitut and English by Inuit adolescents and young adults.

If we now return to Nunavik—a region in which, generally speaking, Inuktitut use is more frequent than it is in Nunavut—we can put the apparent rise in the use of English among young people into perspective. Given what we have just seen in Nunavut, it is not clear that such a rise in the use of English in Nunavik, which is occurring especially among adolescents and especially in the larger settlements, means that Inuktitut is threatened there. Though English is, admittedly, associated with both the colonial history of the region and with globalization, its use is nevertheless restricted by informal means, such as those reflected in the use of Inuktitut between parents and children, as well as by institutional means, such as Inuktitut language policies, education, and the official use of Inuktitut in Inuit governing bodies. Together, such policies and practices have created a place for Inuktitut not only in modern institutional settings but also in community life.

Given what I have described in this chapter, language choice in face-to-face interaction in Nunavik and Nunavut clearly involves complex sociolinguistic processes. Political, economic, cultural, and social-psychological factors all come together to shape language patterns, which have important consequences for Inuktitut language maintenance. Ethnicity, identity, and the meaning of "being Inuk" can no be longer be defined solely in terms of one's tendency to speak Inuktitut. In other words, it has become necessary to recognize the emergence of bilingual Inuit identities. Moreover, since the importance of maintaining Inuktitut is widely acknowledged in Arctic communities, and the political will to do so is similarly widespread, these bilingual identities may remain a feature of Arctic communities for many generations to come.

CONCLUSION

In this chapter, I have tried to show how both English and Inuktitut have come to play pivotal roles in the contemporary construction of Inuit identities in the eastern Canadian Arctic. Although these languages compete with each other in much of the Arctic, and with each other and French in northern Quebec, and although there has historically been an imbalance in the political and economic power held by dominant and indigenous languages in this region, each language is currently seen as crucial to Inuit goals of modernization, self-sufficiency, and autonomy within the Canadian nation-state. Each language is also crucial in the construction of (new) group identities. This is because these identities reflect fluid social categories—including those identified as Inuit, Francophone, and Anglophone—which may involve members of the same social or ethnic group or members of different groups, and which are constructed in part by language practices. And just as new identities are emerging, new language practices are enacted daily—practices that may eventually give rise to new language varieties and new ways of defining Aboriginality (such as those described in Darnell 2004). The nature of this fluidity is revealed, at least to some extent, in the interview and ethnographic data that I have presented here. What is even more clearly revealed by these data, and by my description of Nunavut and Nunavik more generally, is the intriguingly complex role that English plays there in constructing ethnicities and Canadian Inuit identities.

NOTES

1. An exception to this has been the research on Aboriginality and English reported in Darnell 2004.
2. Nunavik leaders are currently negotiating with Quebec for increased self government, which could lead to higher levels of co-management and increased profit-sharing opportunities for Inuit in economic development projects.
3. For instance, in the March 2004 election for Nunavut Premier, one of the leadership candidates, Tagak Curely, refused to use English in his interviews with reporters, relying solely on Inuktitut. Curley lost the leadership race to incumbent Nunavut Premier, Paul Okalik.

REFERENCES

Barth, Fredrik. 1969. *Ethnic Groups and Boundaries: The Social Organization of Culture Difference*. Boston: Little, Brown and Company.

Bunte, Pamela and Robert Franklin. 2001. Language revitalization in the San Juan Paiute Community and the role of a Paiute constitution. In *The Green Book of Language Revitalization in Practice*, Leanne Hinton and Kenneth Hale (eds.). San Diego: Academic Press. 255–264.

Burnaby, Barbara. 2002. Language policies in Canada. In *Language Policies in Education: Critical Issues*, J Tollefson (ed.). Mahwah, NJ: Lawrence Erlbaum. 65–86.

Creery, Ian. 1994. The Inuit (Eskimo) of Canada. In *Polar Peoples' Self-Determination and Development*, Minority Rights Group. London: Minority Rights Publications.

Crystal, David. 2000. *Language Death*. Cambridge: Cambridge University Press.

Darnell, Regna. 2004. Revitalization and retention of First Nations languages in southwestern Ontario. In *Language Rights and Language Survival: Sociocultural and Sociolinguistic Perspectives*, Jane Freeland and Donna Patrick (eds.). Manchester: St. Jerome Publishing. 87–102.

Dickason, Olive P. 2003. Metis. In *Aboriginal Peoples of Canada: A Short Introduction*, Paul Robert Magocsi (ed.). Toronto: University of Toronto Press. 189–213.

Dorais, Louis-Jacques and Susan Sammons. 2000. Discourse and identity in the Baffin region. *Arctic Anthropology* 37, 2: 92–110.

Dorais, Louis-Jacques and Susan Sammons. 2002. *Language in Nunavut: Discourse and Identity in the Baffin Region*. Iqaluit: Nunavut Arctic College; Québec: GÉTIC (Univeristé Laval).

Dorian, Nancy. 1989. *Investigating Obsolescence: Studies in Language Contraction and Death*. Cambridge: Cambridge University Press.

Fishman, Joshua. 2001. *Can Threatened Languages Be Saved? Reversing Language Shift Revisited: A 21st Century Perspective*. Clevedon: Multilingual Matters.

Francis, Daniel and Toby Morantz. 1983. *Partners in Furs: A History of the Fur Trade in Eastern James Bay 1600–1870*. Kingston/Montreal: McGill-Queen's University Press.

Gilroy, Paul. 1987. *There Ain't No Black in the Union Jack*. London: Hutchison.

Grenoble, Lenore and Lindsay Whaley (eds.). 1998. *Endangered Languages: Current Issues and Future Prospects*. Cambridge: Cambridge University Press.

Hall, Stuart. 1992. New ethnicities. In *"Race," Culture and Difference*, James Donald and Ali Rattansi (eds.). London: Sage.

Harper, Kenn. 1983. Writing in Inuktitut: An historical perspective. *Inuktitut*, September 1983, 3–35. Ottawa: Indian and Northern Affairs. Also available on the National Library of Canada website, www.nlc-bnc.ca.

Hensel, Chase. 1996. *Telling Our Selves: Ethnicity and Discourse in Southwestern Alaska*. New York/Oxford: Oxford University Press.

Kinkade, Dale. 1991. The decline of native languages in Canada. In *Endangered Languages*, R.H. Robins and E.M. Uhlenbeck (eds.). Oxford: Berg. 157–176.

Kusugak, Jose. 2000. *Self-Determination in Nunavut: Inuit Regain Control of Their Lands and Their Lives*. IWGIA. 20–23.

Marsh, Rev. Donald Ben. 1964. History of the Anglican Church in Northern Quebec and Ungava. In *Le Nouveau-Québec: Contribution á l'étude de l'occupation humaine*, Jean Marlaurie and Jacques Rousseau (eds.). Paris: Mouton. 427.

Meeuwis, Michael. 2000. Flemish nationalism in the Belgian Congo versus Zairian anti-imperialism: Continuity and discontinuity in language ideological debates. In *Language Ideological Debates*, Jan Blommaert (ed.). Berlin: Mouton de Gruyter. 381–423.

Nunavik Commission. 2001. *Amiqqaaluta: Let Us Share. Mapping the Road toward a Government for Nunavik.* Report of the Nunavik Commission, March.

Olson, David. 1986. Intelligence and literacy: The relationships between intelligence and the technologies of representation and communication. In *Practical Intelligence: Nature and Origins of Competence in the Everyday*, Robert J. Sternberg and Richard K. Wagner (eds.). Cambridge: Cambridge University Press. 338–360.

Olson, David, Nancy Torrance, and Angela Hildyard. 1985. *Literacy, Language, and Learning.* Cambridge: Cambridge University Press.

Patrick, Donna. 1994. Minority language education and social context. *Etudes/Inuit/Studies* 18, 12: 183–199.

Patrick, Donna. 2001. Languages of state and social categorization in an Arctic Quebec community. In *Voices of Authority: Education and Linguistic Difference*, Monica Heller and Marilyn Martin-Jones (eds.). Westport CT: Ablex. 297–314.

Patrick, Donna. 2003a. Language socialization and second language acquisition in a multilingual Arctic Québec community. In *Language Socialization in Bi- and Multilingual Societies*, Robert Bayley and Sandra Schecter (ed.). Clevedon: Multilingual Matters. 165–181.

Patrick, Donna. 2003b. *Language, Politics, and Social Interaction in an Inuit Community.* Berlin/New York: Mouton de Gruyter Press.

Patrick, D. and P. Shaerwood. 1999. The roots of Inuktitut bilingual education. *The Canadian Journal for Native Studies* 19: 249–262.

Rampton, Ben. 1995. *Crossing: Language and Ethnicity among Adolescents.* London: Longman.

Shearwood, Perry. 2001. Inuit identity and literacy in a Nunavut community. *Etudes/Inuit/Studies* 25, 1–2: 295–307.

Street, Brian. 1984. *Literacy in Theory and Practice.* Cambridge: Cambridge University Press.

Wolf, Eric. 1982. *Europe and the People without History.* Berkeley: University of California Press.

8

CONSTRUCTING A DIASPORA IDENTITY IN ENGLISH: THE CASE OF SRI LANKAN TAMILS*

A. Suresh Canagarajah

Sri Lankan Tamils are among the new wave of immigrants in Western metropolises, such as London, Los Angeles, Paris, Sydney, and Toronto. Before the Sri Lankan ethnic conflict was really militarized (i.e., before 1983), Tamils left the island sporadically for educational purposes, and largely to the colonial metropolis London. Within the last twenty years, however, they have spread out in increasing numbers to many other locations in the West. Currently, according to the estimates of Tamil community organizations, there are about 150,000 Tamils in Toronto (the largest community of Tamils outside Sri Lanka) and 50,000 in London. Though they are scattered in the United States, Tamils form a cohesive community of about thirty-five families in a small town populated by 121,000 people in Lancaster (about sixty miles north of Los Angeles).

I have been visiting these three locations (i.e., Toronto, London, and Lancaster) since 1996 to understand the Tamil community's orientation to the English language. In addition to extended periods of stay in each location for ethnographic purposes, I have also been conducting sociolinguistic surveys to understand patterns of bilingualism and interviewing families to explore their attitudes to language. A point that parents and elders kept repeating in all three cities was that Tamil language was going to "die" within the community in the West and that the Tamil identity may get erased within the next fifty years or so. They also compared themselves with the Indian, Chinese, and Hispanic communities in the West, and asked how it is that they could sustain their bilingualism for generations, whereas the

Tamils are shifting so quickly toward English monolingualism. The results from the questionnaires and ethnographic observation confirm their impression about language shift. Whereas the grandparents are largely monolingual in Tamil, and the parents are bilingual, the children are overwhelmingly monolingual in English. Furthermore, a large number of parents are speaking to their children only in English. Similarly, children are speaking to each other in English within their family and across families. Despite this fact, the youth who spoke to me said that they think of themselves as Tamils and don't feel that their Tamil identity was going to get erased in the future. However, the elders are of the view that Tamil identity and community are impossible without the Tamil language. I wish to explore in this essay the new relationships that are emerging between language and community in the Tamil diaspora in particular, and in postmodern communities in general. Is it possible for the Tamil language to die, but Tamil identity and community life to remain vibrant?

BACKGROUND

Before I go into that question, it is important to orientate to the history of the Sri Lankan Tamil community in order to appreciate the ironies in the new linguistic developments. When the British controlled the island from 1796 to 1948, it was religion that was the core value for Tamil-speaking people. Nationalistic thinkers of that time made clear that learning English was acceptable as long as people didn't convert to Christianity (see Canagarajah 1999). To assist in this process, local leaders even built Hindu schools for providing an English education. But when the British left the island in 1948, the minority Tamil community faced a different challenge. The dominant Sinhalese community declared Sinhala the official language. With this legislation, language became a core value for Tamils around this time. When a democratic struggle for language rights turned out to be unsuccessful, community leaders sought autonomy for Tamil regions. As election mandates to this effect were not honored, around the early 1980s the youth launched an armed struggle for a separate state named "Tamil Eelam." When the Sri Lankan state unleashed its military power to keep the Tamil regions under its control, the Tamil people started fleeing the island in order to escape the increasingly intense warfare and the periodic revenge attacks on civilian lives and property by Sinhalese soldiers. The people who have been coming to Toronto, London, and Lancaster are those with a heightened linguistic consciousness and community solidarity. In fact, Tamils have traditionally been a very homogeneous and conservative community, rooted in their homeland for generations. It is therefore intriguing to many elders in the community that such people should suddenly make a wholesale shift to monolingualism in English.

SOCIOHISTORICAL FACTORS IN LANGUAGE SHIFT

Before I present some of the emerging patterns of linguistic interaction and community membership, let me explain the factors that may be precipitating the shift toward English away from Tamil.

Many community elders pointed out to me that they are unable to visit the homeland to give their children an education in Tamil or an exposure to Tamil lifestyle. They feel that it is the ability of the Indian, Chinese, and Hispanic communities to have a more effective transnational life that permits them to remain bilingual. It is true that the ongoing warfare has prevented Tamils from visiting Sri Lanka. Many areas of the traditional Tamil homeland are out of bounds and under tight security, shifting in control between the Tamil militants and state forces periodically. Also, many respondents stated that they didn't have a sense of patriotism toward Sri Lanka. The history of animosity with the Sinhalese community has prevented any sentiment of shared nationhood from developing. Perhaps all this has encouraged Tamils to focus on their new life in the West rather than hanker after their past in Sri Lanka.

Furthermore, the direction taken by youth militancy has soured people's attitudes toward the nationalistic struggle in their homeland. After internecine fighting among Tamil groups, causing unnecessary bloodshed, the group that is dominant now—the Liberation Tigers of Tamil Eelam or LTTE—has imposed an authoritarian rule in the territory under its control. Children are sometimes forcibly recruited to join the militia. The de facto regime has also mandated a linguistic policy of Tamil Only for official or formal interactions. Sanskrit and English borrowings are excised from Tamil, and new words are being coined for those purposes. Many Tamils find these activities too chauvinistic and narrow for their tastes (see Canagarajah 1995).

The legal restrictions in their new communities of habitation also prevent Tamils from freely shuttling between their homeland and the West. Those who reach the West as political refugees destroy their travel documents before they claim asylum. Till they are officially granted resident status and new travel documents by the host country (which can sometimes take ten to fifteen years), they cannot travel anywhere. Also, having stated that they are fleeing Sri Lanka because of inhospitable living conditions, they will contradict themselves if they make frequent trips there for holidays.

In addition to these external restrictions, there are also some community internal factors that affect language shift. Our colonial connection with Britain has created widespread proficiency in English across generations among the middle class. Thus many children lose the positive influence of their grandparents in maintaining the vernacular. There are studies that show that among migrant communities in the West the grandparent–grandchild relationship plays a favorable role in language maintenance (Boyd and Latomaa 1999). In the case of the Tamil community, both the parents and some grandparents are proficient in English, and they switch to English when they converse with children born here. There is an interesting point of contrast on this matter with Tamil families in Paris and Germany. Since the parents don't have prior exposure to French or German in their homeland, they stick to Tamil in their interactions with their children at home, even though the children are fluent in the host languages. Because of this, Tamil children in non-English speaking countries have a better proficiency in the vernacular. Thus our colonial legacy seems to accelerate our shift toward monolingualism in English now.

The lower-middle-class families, which have had less proficiency in English in the homeland, display other reasons for deliberately pushing their children to develop proficiency in English than in Tamil. Some of these families have to depend on their children to conduct institutional relationships in the host community. Children accompany the parents to negotiate with immigration centers, law courts, and schools. The parents themselves begin to learn English when they arrive here, and hold their host community accents as a badge of honor against the traditionally bilingual elite (the educated middle class from Sri Lanka) who still speak standard Sri Lankan English. Many underprivileged families are seen to be making up for their status loss at home with the new opportunities they have in the West for learning English. In my interviews with them, parents from such backgrounds tended to excuse their children for not knowing Tamil. Those from more privileged backgrounds were, on the other hand, more apologetic about their children's lack of proficiency in Tamil. Some respondents were prepared to confess that their attitude reflects the prestige value that English has gained since colonial times in the Tamil community. They commonly referred to this attitude as *aankila mookam* ("the craze for English").

Just as the less privileged families used English to construct new identities in the new land, female members of the community also discover similar benefits. In the questionnaire answered by the youth, more girls than boys mentioned that it was not necessary to use Tamil in their host community (e.g., asked whether it was important to use Tamil at home for communication, seventy percent of the girls disagreed, whereas only fifty-eight percent of the boys responded negatively.) In the interviews, whereas boys were apologetic about their lack of knowledge of Tamil, the girls were frank about the importance of English. Boys made statements like the following: "It is important to know Tamil, but we don't have time to learn it"; "It is good to know Tamil to display our culture, but it is not of practical value here"; "I would have liked to learn Tamil, but my parents didn't take the time to teach me." In the case of girls, they said: "This is London, not Jaffna. What's the point of talking in Tamil?" or "Tamil will only keep us down as a minority" or "We should keep back our own languages and speak one common language if we are to join the mainstream life here." There is widespread anecdotal evidence that more women than men switch to English when they meet another Tamil in a public place in London, Toronto, or Lancaster. It is possible that Tamil females are enjoying a new sense of freedom and individuality that women haven't experienced traditionally in the Tamil community. Perhaps women are taking on to English more enthusiastically as it provides alternate identities that favor their interests in the new life in the West.

COMMUNITY FOR TAMIL YOUTH

In the context of the above macrosocial factors, I like to take a closer look at the linguistic life of Tamil youth—here defined as those under twenty-eight years old and not married. The youth in Lancaster overwhelmingly replied in my survey questionnaire that English was their dominant language.[1] Only 1 among 22 respondents said that she was a balanced bilingual in English and Tamil. The others had only passive competence in Tamil, if at all. All of them said that they used solely English for communication with their siblings. All except 1 said that they use English with their parents. The only respondent who answered differently said that she uses both languages. Since she is only 8 years old, it is not clear whether she'll sustain this bilingualism as she grows up. Of their parents, more than half the number said that they use only English with their children. About twenty-five percent used both languages, a similar number (mainly mothers) said they used mostly Tamil with their children. Both latter groups said their children responded solely in English. Such communication is possible because the parents are overwhelmingly bilingual. Whereas the grandparents I interviewed were all monolingual in Tamil, 17 out of the 22 youth I interviewed said that they responded to their grandparents in English. Apparently the youth used the nonverbal aides of gesture, context, and tone to assist their communication with their non-English-speaking grandparents.

The language attitudes of the youth also show them favoring monolingualism in English. Only 2 out of 22 respondents said that it was important for their generation to be proficient in Tamil. Even the 2 who differed from the rest agreed with the majority in saying that it was not important for them to be literate in Tamil. All except 1 said that Tamil will soon run out of use among those living in the West. There was however a difference of opinion relating to whether it was important for families to use Tamil at home. Out of 22 youth 8 said that it was important to do so. Parents also show a similar interest. Only 2 out of the 27 who responded said that it was not important to use Tamil at home. This desire to have some Tamil spoken at home may suggest an attempt to maintain some form of Tamil identity at least in limited contexts. However, about 18 out of the 22 parents said it was not necessary for their children to be *literate* in Tamil. The parents have probably lowered their expectations over time. Apparently, for the children to at least speak informally at home seems to suffice for the parents. In fact, 24 out of 27 respondents agreed that Tamil was in danger of dying out in United States.

Many institutions in the community are beginning to acknowledge this monolingual reality among the youth. Ethnic churches in London and Toronto now worship bilingually. The original intention of these churches was to offer worship in the vernacular in order to build community solidarity. Now, in the churches I visited, there is simultaneous translation of everything that transpires during worship. The clergy acknowledge that the youth cannot follow the sermons or prayers in Tamil. In fact, Sunday schools for these children are held solely in English. Furthermore local literati, who have published magazines and newspapers in Tamil for some time, are now turning their attention to publishing developments in vernacular literature and culture in English. The feeling is that they can develop in the younger generation an awareness of Tamil history and culture at least by communicating in English. New journals are already being published in English in the Internet to fulfill this need. A Tamil teacher I spoke to said that he has written to Tamil newspapers in Toronto to have a few pages in English so that the youth will be drawn to reading these journals. He felt that though Tamil language and culture may be lost, at least community life will be preserved through these publications.

Despite this picture of a pronounced shift to monolingualism among the youth, their sense of identity is more complex. It was in the more relaxed settings of my oral interviews with them that they were prepared to contemplate their attitudes to identity and community. Asked in general if they would identify themselves as Tamils, the youth said that this identity was irrelevant now. They declared that they would identify themselves as British or Canadian or American (as the case may be). But when the extreme scenario of a Tamil identity or community being completely wiped out in the next generation was posed, they considered this improbable. They felt that there would still be a sense of Tamilness that constituted their identity.

I have explored with several youth groups what would constitute their sense of Tamilness. A majority of them agreed that their Tamil identity would *not* be based on some of the traditional markers that have been important for their parents: that is, they may not speak Tamil language; they may have lost their faith in Hinduism; they may not identify with many of the cultural values important for the community—such as suppressing individual aspirations for the sake of collective good, obeying the parents unconditionally, or going along with an arranged marriage based on caste and dowry; they may not eat Tamil food all the time; they may not dress in the traditional sari or verti; women may not wear their hair long, and men their hair short; they had no desire to go back to the homeland; and they didn't share the political ideology of Tamil autonomy or linguistic/cultural purism. In fact, they surprised me by questioning the exclusivist assumptions of my survey questions. They argued that identity should not be based on all-or-nothing constructs—that is, American or Tamil. For them, identity should not be based on displaying a finite set of distinguishing features. They all argued that Tamil identity has to be defined differently. They made statements like the following: "Identity is very flexible"; "Identity needs to be defined more broadly"; "Being a Tamil shouldn't be based on displaying traditional norms."

Their statements on what alternative constructs this identity and community life will be based upon initially sounded vague. Some simply said that they thought of themselves as Tamils—"in the heart of hearts, we know we are Tamils" or "We feel we are Tamils." The heightened political conflict in the homeland, the constant flow of Tamils to the West, and the controversial media discourse on Tamil refugees kept a Tamil identity constantly alive before them. Also their

skin color and physiognomy, added to experiences of racism in schools
and in the wider society, always reminded them that they were different. During the time of my fieldwork, I thought of constructs like
thinking and feeling as too flimsy to work with. But Arjun Appadurai
(1996) and other postmodern anthropologists have developed
Benedict Anderson's notion of nations being imagined into being
from print media. They theorize how imagination thus plays an
important role in building a sense of community. Appadurai states:
"The many displaced, deterritorialized, and transient populations that
constitute today's ethnoscapes are engaged in the construction of
locality, as a structure of feeling, often in the face of the erosion, dispersal, and implosion of neighborhoods as coherent social forma-
tions" (1996, 199). "Structure of feeling" is certainly a construct that
sounds nebulous compared to the more solid empirically tested
traditional constructs such as core values, cultural practices, social
networks, and community solidarity. But we realize that in postmodern cultural life affective realities constituted by discourses and semiotic codes are powerful enough for "locality" or community life to be
built upon.

Another suggestion proffered by the youth for the Tamil identity
they enjoy despite losing the traditional markers of community is
more interesting. They said that the family activities they still participate in would provide them a Tamil identity. A teenager from
California said, "As long as we have our Tamil families here, we will
have a sense of being Tamils." The youth envisioned continuing to
participate in family-oriented life in the West. There is a special reason
why family life will form the basis of Tamil community life in the diaspora. An officer in a community service organization in Toronto
pointed out that about ninety percent of the Tamil population in
Canada belong to "cluster families." What he meant was that the
Tamils who migrated here belong largely to extended families. This is
so because the governments in these countries of refuge provide special immigration privileges to the siblings and parents of someone who
is already here. It is for this reason that the youth felt that family life
will be central in their life in the West. They will continue to attend
traditional functions, ceremonies, and events and interact with other
Tamils in these family gatherings. What would define them as Tamils
is their participation in these family gatherings and the knowledge of
family practices that will prove them as insiders. In saying this, the
youth placed emphasis on practice rather than knowledge, culture,
discourse, or ideology as defining identity. They could adopt different
values and beliefs, but they could still meet with other Tamils on

specific occasions to do things together as a community and enjoy this identity for however limited duration and settings in their everyday life.

To illustrate this point, let me take the event of attending a pooja or bhajan in someone's house. The youth who accompany their parents every Friday to a relative's house may in fact be atheists, agnostics, or nominal Hindus. They may not have the language to understand or respond to the forms of worship. What is important is that they still know what is involved in this social activity. They can participate in the different rituals and stages of the pooja. In fact, language was never the issue in the Hindu pooja, as no one ever understood the Sanskritized mantras of the Brahmins anyway. What was always important was simply the participation in an activity by a community of believers. Though the youth I spoke to didn't articulate it precisely in this way, I find it useful to think that they are defining their identity as deriving from their participation in a community of practice.

Parents, in fact, seem to be working on this assumption as well. They now transliterate Tamil hymns and prayers in English to be memorized by their children and recited in their religious gatherings. These parents seem to acknowledge that meaning or believing what one says is immaterial. What is more important is participation in an activity with others in the community. There are other parents who don't mind if their children don't speak Tamil, but still send their children regularly to traditional dance or music classes. Here, the classes are held in English. Even the transcriptions/notations/lyrics of these traditional art forms are written down in English. Though some older members of the community find it bizarre to teach these arts in a foreign language or without an understanding of the cultural/linguistic implications, what is important for the parents and children is the ability to be able to practice these art forms. This is enough to qualify as Tamil identity of community membership for many. This attitude even influences Tamil language classes. In London, I observed classes for teaching Tamil held on weekends in public schools. Before classes, students spoke in English in the corridors. Teachers spoke in English when the classes started. They adopted a product-oriented pedagogy of teaching lexical items, grammar, and syntax. Communicative competence didn't seem to be important. For these teachers and students (and parents who send the children here) what is important is the rudiments of the native language to signal community membership or to aid the children in following in group practices. In fact, going to Tamil classes itself seemed to suffice for many. Interacting with other

Tamil kids and enjoying some community interaction seemed to be the point behind this weekly exercise for many. Parents too would socialize among themselves when they drop their kids for classes.

The notion community of practice is gaining importance in educational circles today. After attempting to explain disciplinary communities in terms of unifying/homogeneous discourses, we now acknowledge that there is often nothing in common between members in a specific community who hold multiple group memberships, diverse identities, and divergent beliefs. What holds the community together is the fact that members come together with others sharing similar objectives to engage in specific projects and purposes (see Prior 1998). It is practice rather than beliefs or identities that gives them a tentative unity and coherence. Lave and Wenger (1991), who use this concept to explain the socialization of apprentices into diverse professions, emphasize practice for two different reasons.[2] Not only is it the ability to perform the practices that constitute that community which confers membership for someone; this competence is also acquired though practice. One cannot be taught such competence formally, theoretically, or deductively; it has to be acquired by doing the different types of activity that constitute this community life in a nurturing company of experts. Furthermore, discourse is a byproduct, not an essential feature in this process of professionalization. Apprentices gain the needed discourse in context, and learn to work with a range of divergent discourses that will find relevance in the activities of the community.

As diaspora and transnational communities are becoming increasingly complex in constitution, and community membership is becoming decentered and deterritorialized in postmodern contexts, scholars are searching for new ways of defining identity. The Tamil youth, for example, are not sure if they can identify themselves as belonging to one community or the other—or even if they constitute a new community. They simply shuttle between different communities of practice to adopt contextually relevant and strategic identities. They may even find the conflicts and differences between communities to their advantage as they provide options for them to construct newer identities. It is also important to recognize the desire of individuals to separate themselves from communities to adopt alternate positions. As Homi Bhaba (1994) has theorized, the "in-betweenness" of postcolonials is of strategic importance. Postcolonial subjects find different forms of identity between communities, discourses, and subject positions. From this point of view, it is difficult for Tamil youth to give a straightforward affiliation with one community or the other. Their identities are constantly in the process of making, as they shuttle between different communities. Not defining themselves according to set identities is of additional advantage as they can strategically adopt favorable identities depending on the context.

If such are the preferred modes of subjectivity, they have serious implications for the coherence of communities. Communities made up of such shifting subjects become so fragmented, multiple, and transitory that Papastergiadis (2000) asks whether we can define communities without resorting to "essences":

> Can there be communities without the guarantees of stability? Is the essence of a common language and shared history the only guarantee for a collective identity? . . . Communities overlap, abut, and adjoin each other. What holds them together can rarely be identified by unique values or an exclusive set of characteristics (196–197).

To answer such questions, Appadurai (1996) and Papastergiadis (2000) argue that social sciences may have to adopt a fundamental shift in frameworks. Both scholars urge that their discipline should give up mechanistic and static models and adopt more fluid metaphors from fields such as chaos theory (Appadurai) or aesthetics (Papastergiadis). In fact, both find the metaphor of "turbulence" fascinating as it accommodates the fluidity of identities and community membership while also symbolizing the tentative patterns and harmonies that emerge. Papastergiadis says: "We need to explode the myth of pure and autonomous communities, reject the earlier mechanistic and territorial models of community and present new perspectives on the concepts of space and time which can address the dynamic flows that make community life. There is a need to take a more processual view of power and agency, to note that communities are not just dominated by rigid structures and fixed boundaries but are like a 'happening' " (2000, 200). The framework I adopt here is less mystical. For me, the communities of practice construct provides one way in which we can explain the notion of a community identity developing among Tamil youth. People enjoy membership in a community not because of what they believe in, who they are, or what they know, but because they come together to do certain activities. It is in this sense that it is possible to belong to the Tamil community without knowing the Tamil language or sharing everything in the Tamil culture.[3]

NEGOTIATING DIASPORA IDENTITIES IN ENGLISH

Defining one's identity as based on membership in diverse communities of practice provides considerable flexibility for individuals to enjoy multiple identities in a contextually relevant manner in shifting relationships. Tamil youth move in and out of communities, adopting different—sometimes conflicting—identities if it suits their purpose. Language and discourse mark their shifting relationships and aid their shuttling activity, although they don't determine their identities.

Let me illustrate the linguistic implications of communities of practice by narrating my observation of children from an extended family who spent a summer vacation together in Toronto. I had the opportunity to live in the same house and observe their interactions. The subjects were all cousins, consisting of three from Toronto, two from London, and two from the United States, ranging in age from nine to nineteen years. (I refer to them hereafter as "cousins.") Though they spoke the English that approximated the "native" dialects of the locations they came from, they had little problem understanding each other. Their parents of course complained that sometimes they couldn't understand their nieces or nephews from another country. But the cousins were apparently multi-dialectal in English. They also had the passive competence to understand the Sri Lankan English of their uncles and aunts. Occasionally they would parody the Sri Lankan English of their parents or relatives. As for their own dialects, they maintained them to symbolize their distinct identities without shifting to another. They took pride in the dialects they spoke. There were no jokes about each other's English.

Beyond their dialects, it was clear that the cousins shared a range of discourses that their parents didn't know. They often engaged in talk about the Internet, pop music, cinema, fashion, and entertainment. These discourses were commonly shared by them so that they could communicate without problems. But here, again, their parents were at a disadvantage. Their discourses marked their membership in the global pop or teen culture that their parents didn't belong to. These discourses also marked their engagement in other youth groups in their respective countries where they met to perform other activities together. To engage in such activities and enjoy those memberships, the Canadian hosts would occasionally take their cousins to the mall or the cinema where they hung out with Canadian friends and engaged in shared discourses. Thus they would sneak out of their houses everyday to interact with non–Sri Lankan friends from work or school. They had a community beyond the Sri Lankan family with which they shared other activities. When they left the home, their accents and discourses changed. They also dressed differently. Even their names changed. The cousins stopped calling each other Suja and Suren (i.e., their Tamil names), and adopted their English names Joanna and Jonathan.

But when it came to family-oriented events and activities, the cousins participated in them as well without problems. In prayer meetings at home, birthday parties, and weddings (which are some of the gatherings I observed), they mingled with the older members. They would help around in getting the house ready for these events, with an insider awareness of the rituals. There was no visible displeasure in participating in these events. They seemed to fluidly move from one community/context to the other. Their language, however, showed some difference in this context. They would mix a lot more Tamil words in their English when they interacted with family members. Words for food, clothing, and relationships (*ammaa*, mother; *appaa*, father; *akkaa*, older sister; *tambi*, younger brother; *maamaa*, uncle) feature more often in their English in these contexts. (Such words are censored when they interact with their local friends outside the house.) This code-mixed variety of English was enough for them to symbolize their ingroup status in this Tamil community of practice. That their bilingual parents engaged in code switching helped the children to follow what was going on. Also some of the older children had passive competence in Tamil to understand the flow of conversations and interactions. More importantly, because they were insiders to these practices, the cousins didn't depend on language for their participation.

Consider, for example, how the communicative event called "prayer meeting" would happen in this community. Because religion is a very important part of the family life of Tamils, both Christians and Hindus will get together at least on a weekly basis with their extended family for prayers. Sometimes this is a more public event, where people gather with many other families in someone's house for worship. In one such event I observed, the cousins joined their parents and grandparents in an evening of Christian prayer. Though the cousins were initially watching an English video in the basement of the house, they quickly came upstairs for the prayer meeting when one of the parents called out to them. There was no visible display of dislike for participating in the prayer meeting. The cousins first helped set up the living room by rearranging the chairs into a circle where everyone could see each other. The event opened with the singing of some hymns in Tamil. The parents could read the hymns from their books. Though the cousins couldn't read them, they kept silent so as to not disturb the singing. One of them played the piano for the hymn, as the cousins knew the tunes from having heard them being sung before. Then when others closed their eyes and bowed their heads, the cousins knew it was time for prayer—though they didn't necessarily follow the prayer said in Tamil. After that, those gathered sang a medley of short choruses. Some of the choruses had both English and Tamil versions. The cousins sang the English versions. They seemed to even know the Tamil version in some cases, having heard them repeatedly in previous meetings. After that was the reading of the Bible. Since this was in Tamil, the cousins did not understand the passage read. But some of them turned to the same passage in their English Bible, after inquiring from their parents information relating to the chapter and verse. Soon after that, one of the elders spoke about the passage, applying the interpretation to their personal lives. The speaker often code switched. He would first utter a few statements in Tamil, and then say something brief in English that either summarized his statements or made a new point related to what he had already said. Because of the interspersed English, the cousins managed to infer the meaning of the Bible passage. The meeting ended with another hymn and a prayer in Tamil for which the cousins remained silent. When the meeting was over, they asked me to take them to the mall where they would participate with some of their Canadian friends with other communicative events such as sampling newly released pop song CDs in the HMV store.

Though the cousins moved easily from one community to the other to evoke different identities, in many cases they took care not to let these communities overlap or clash. So I would always drop off the kids in the mall or cinema with their non–Sri Lankan friends, but I was never formally introduced to them. The consequences of letting these communities meet are illustrated by an argument between a mother and a teenage son in London while I was interviewing them:

> *Son:* I sometimes don't like *Amma* ["mother"] to come and talk to me in front of my friends when she picks me up from school.
> *Mother:* hm, I always stand away from you and wait for you to come.
> *S:* But you spoke to me in front of Michael yesterday.
> *M:* uh, but I spoke to you in English.
> *S:* It is not about that. It is about, it is about the way you speak. It is like, their parents don't speak like that.
> *M:* Well, I wasn't scolding you or anything.
> *S:* No, but you were saying things like, "come here, go there, do this, do that." Their parents talk more nicely (voice breaks; close to tears).
> *M:* I don't understand what you are talking about (exasperated).

It is clear that what the son is complaining about is not the choice of language (i.e., English or Tamil), but the way the language is used.

He is complaining about tone, attitude, demeanor, pitch, and voice. It is clear from the example that it is not language but discourse that indicates identities and community membership for this young man. From this point of view, this argument is about practice. For this son, the child–mother relationship is practiced differently in the British families; his mother's discourse suggests a different practice. These divergent discourses symbolize different community memberships, and cause discomfort to the son.

There are some contexts, however, where the youth find it strategic to flaunt their Tamil community membership as they engage in other community relationships. Some boys in London told me that they would occasionally use Tamil lexical items they knew—such as terms of friendship/endearment, terms for calling/hailing someone, and profanities—in front of British teachers and students in their schools. They told me that they did this to outsmart some of their British friends, to laugh behind their back, or simply to enjoy a measure of solidarity when they felt insecure. Though they desired membership in their peer groups in the school, they occasionally experienced biases there. So it was important for them to hold this level of solidarity in reserve as an option when they needed it. When they were ostracized by other students, they had their Tamil community membership to fall back upon. Those who were a bit more proficient, shared secrets among themselves within earshot of their British teachers and peers. They even used Tamil to cheat in tests or assignments. In more utilitarian terms, some found the Tamil identity useful in the context of the equal opportunity and diversity discourses in the West. Identifying themselves as Tamils in these contexts would gain them some covert prestige and educational and professional opportunities. It is evident that a form of elite bilingualism is developing among some of the youth. At a late stage, at high school or college level, some are making attempts to develop more proficiency in Tamil.

To understand that these shifting allegiances are strategic, we have to note that in the context of their families these youth invoked their Western community membership. When parents were seen as stifling their interests or controlling their freedom, the youth would simply say, "I am not Sri Lankan. I am a Canadian." They would remind their parents that they simply didn't know how some of these activities and concerns were practiced in the Western community. They would mock them for being too insular or unsophisticated, and not knowing the mainstream ways of behavior and discourse. Through this strategy, they gain a measure of freedom against the age-based status of their parents and elders. Some parents in fact complained of being cheated by their children. The children would give false reasons or excuses for staying late from school or hanging out with their friends, as their parents were not well informed about institutional practices here. Community leaders have also started talking about role reversals in the Tamil community—with youth versed in the Western ways enjoying more power in certain social contexts where the parents are ill-informed.

IMPLICATIONS: NEW ETHNICITIES, NEW DIASPORAS

Given the relationships we see among the Tamil youth, it is clear that the next generation of Tamils outside Sri Lanka will interact primarily in English. Although I mentioned only the interactions of youth from Britain, Canada, and United States, I have also observed interactions between other Tamil youth from France, Germany, Sweden, and the United Arab Emirates in English. Even parents are finding that English is becoming the default language for interactions when they meet across generations from different localities of migration. English is fast becoming the language in which diaspora relationships are conducted in the West. How does this reality influence me to answer the questions posed by the elders at the beginning of my study about the future of Tamil language and community? I have to say: yes, Tamil language will probably die in the Tamil diaspora in the West in the next generation; but no, Tamil community and identity won't disappear.

A Tamil community without the Tamil language? The older generation of Tamils is offended by the idea when they hear it put this way. Their attitude reflects the bias that language is or should be a core value of community identity. We have a long history of defining community based on a unique language. Whether in seventeenth-century West or decolonization East, language has played a crucial part in defining nationhood and community membership. What we are finding in postmodern forms of globalization is the declining importance of language in symbolizing community membership. This means that national languages and vernaculars are becoming less significant for diaspora communities. As we talk about deterritorialization of cultures, we have to also begin to discuss deterritorialization of language. English is not a language solely of the British, just as Tamil is not the only language of the Tamils.

Personally, I find these research findings depressing. I would have liked the Tamil community in the West to be more in touch with their homeland. I would have liked them to maintain the vernacular. I would have liked them to confirm the picture we get of fascinating modes of transnationalism in cultural studies as postmodern subjects accommodate the old and the new, traditional and contemporary, in colorful forms of hybridity. The Tamil youth in the diaspora are bland in their monolingualism. If at all, their hybridity operates within a narrow linguistic and cultural framework. But perhaps these are some of the changes developing in what Appadurai (1996) calls the age of the postnational imaginary. Appadurai argues that though people are now showing less interest in identifying themselves in terms of an originary nation, there are new forms of transnational identities developing. After wrongly grouping Tamils as belonging with Serbs, Basques, and Quebecois in still hungering for a territorial identity, Appadurai (1996) states:

> more impressive still are the many oppressed minorities who have suffered displacement and forced diaspora without articulating a strong wish for a nation-state of their own. Armenians in Turkey, Hutu refugees from Burundi who live in urban Tanzania, and Kashmiri Hindus in exile in Delhi are a few examples of how displacement does not always generate the fantasy of state-building. Although many anti-state movements revolve around images of homeland, soil, place, and return from exile, these images reflect the poverty of their (and our) political language rather than the hegemony of territorial nationalism. Put another way, no idiom has yet emerged to capture the collective interest of many groups in translocal solidarities, cross-border mobilizations, and postnational identities. (166)

I would even go to the extent of saying that for diaspora Tamils, as for certain other migrant groups, community is not based on core values any more. As I indicated in the introduction, the core values of Tamils have been changing anyway according to the threats and conflicts they confronted in history. They shifted from religion to language as their core value when they moved from British to Sinhala colonization. But in the current situation, there is a declining emphasis on core values in defining diaspora identity. The youth prefer to enjoy shifting, temporary, fluid identities in less bounded/binding communities of practice.

All this is perhaps good news for English. It is becoming the global language that serves to cement the relationship of many transnational communities that inhabit different cultural and linguistic localities. English is supple enough to accommodate different registers and discourses that go with diverse communities of practice. So the Tamil youth groups are able to orchestrate different dialects to shuttle between different communities—that is, an English mixed with Tamil for home, and an unmixed English characterized by the discourse of pop culture and Internet for wider relationships among peer groups. Similarly, the regionally distinct dialects of British, Canadian, and American dialects give distinct differences in identity. There are also more subtle differences in status that are being negotiated in English within the diaspora community. The standard Sri Lankan English spoken by the educated bilingual elite is treated with contempt by the less-educated Tamils who learnt their English in the West and adopted local accents. But because of their limited educational and professional attainment, the latter group is familiar with a limited range of registers and discourses. For example, they are not familiar with Internet discourse, academic register, or pop culture. This difference enables the traditional Tamil elite to claim higher status. Accent or register—which defines status? This question is negotiated differently in different contexts of interaction.

What are the implications of all this say for the hegemony of English? Is this a new form of colonization of many non-Western communities into a homogeneous culture of globalization? But note that cultural diversity doesn't stop because English is used. Diverse communities of practice, with diverse values, can still function in English. In fact, English is becoming pluralized by these ethnic groups to serve their ingroup purposes. The pluralization of English can be bad news for purist "native" speakers who would bemoan the molding of language into shapes and shades they cannot recognize. This is another way in which English language is becoming deterritorialized. The language is losing its traditional ethnic and territorial identity, as it is now used by communities beyond the West for their own purposes and values (see Canagarajah, 1999).

But there is a more serious resistance politics behind the new identities and communities developing transnationally in English. Stuart Hall (1997) talks about different reactions by migrants in the West to different forms of globalization. In the earlier form of modernist globalization, ethnic groups resorted to returning to their roots—which meant an essentialized and homogeneous identity that accentuated their polar difference from the dominant community. This meant holding on to the vernacular. The reaction to this by agencies of postmodern globalization was to work out a subtle hegemony that accommodated these differences by providing them a secondary place in the status quo. Scholars may even consider the multiculturalist discourse in contemporary society as an example of this form of hegemony. Multiculturalism means that all the groups in the West will keep their differences under control, in deference to the dominant culture and social formation in their new lands of habitation. The strategic response of minority and migrant groups then is to display the complexity of their identity. They are in effect saying that they cannot be stereotyped and objectified. Their identity is shifting, multiple, contextual, and therefore strategic. They refuse to be boxed into one position or the other. They infiltrate mainstream communities, shuttle flexibly, and still celebrate their differences when necessary. This strategic adoption of diverse identities in different contexts to suit one's purpose can be unsettling to the status quo, which would like to assign fixed places for different groups in the new order of things.

Note also that Tamil language is in no danger of dying. The cultural development of the community in the homeland is diametrically opposed to the one we have seen in the diaspora. In the homeland of Tamil Eelam, the people and local regime are working toward linguistic purity and a return to classical premodern values. As long as this development goes on there, it would fuel structures of feeling for the imagination of alternate/local identities for diaspora communities.

However, this contradiction raises questions about the relationship between the diaspora and the homeland. It appears that both communities are moving apart. Ashis Nandy, in an interview with Papastergiadis (1998), argues that it is time now to separate these two communities from their coupling. Based in India, Nandy argues that any attempt on the part of the Indian diaspora to speak for the homeland is futile. Their interests and concerns are very different from each other. So he argues, "Even for the diaspora to be itself, the expatriate Indians must acknowledge that one cannot have it both ways. One cannot be an expatriate and, at the same time, demand to set priorities in the mother-country. Psychologically, this means that one must not just develop cultural links, but also differences" (Papastergiadis 1998, 111).

What this research shows is that there are tensions and conflicts within transnational communities that we have to begin to acknowledge. Papastergiadis (1998) argues that social sciences have sought to develop constructs that are unified and homogeneous despite acknowledging the decentering and deterritorializing work of postmodern globalization. Consider the construct hybrid, for example. We are attempting to fuse the disparate linguistic and cultural influences of transnational communities into a new whole. In this process, differences that are celebrated are again contained or harmonized into a unified whole. But Papastergiadis favors recognizing the unresolved tensions in the diaspora. Many scholars have argued in favor of using hybridity not to contain but to proliferate difference (see Mignolo 2000). There is, accordingly, a search for definitions of community that don't stifle difference in favor of unity. Papastergiadis (2000) prefers the term "cluster" rather than "community." He defines clusters as:

> a space in which various participants gather, and in the process of assembly the respective identity of each member is respected, but at the same time a motion, shape and energy are generated by their proximity. Simultaneously, a semi-porous boundary is formed and new sets of possibilities are established. Within such a space it may be necessary to hold a number of differences together, to arrange them in multi-directional and fluid orders, and, most importantly, not to reduce the identity of one as the negative of the other. . . . To participate in a form of belonging with others may not require that we all feel as one, that we have a common origin, or only speak in the same language. Unlike the dominant narratives of the nation-state, clusters are held together by their *inessential* features. (Papastergiadis 2000, 210)

My use of the notion community of practice to explain the diaspora life of the Tamils is motivated by a similar conceptual orientation.

The case of Sri Lankan Tamils raises questions about ways of defining diaspora communities. Why is it that the Tamils are different from other diaspora communities, such as the Indians, Chinese, and Hispanics, who have a stronger grounding in their native language and culture? Perhaps we have to distinguish between migrant communities to recognize important differences among them without generalizing everyone. Arturo Tosi (1999) argues that the new wave of migrants is less unified, settled, and bounded compared to the earlier waves of immigrants to the West. The people from Africa and Asia who flee violence in their countries nowadays are coming to neighborhoods that are less homogeneous, lacking in support groups, and prone to be stigmatized by the dominant community. Even members of the same community who came in previous waves of migration for professional purposes sometimes stigmatize the new members who come illegally as refugees. Lacking support from their own communities, the new immigrants are forced to fend for themselves and establish institutional and social relationships on their own. Such people may develop their identities in ways that are strategic in the new context, rather than being bound by traditional norms. Tosi can understand if there is even confusion and instability in these groups in relation to their identity and membership.

Perhaps we have to distinguish between different types of diaspora. Robin Cohen (1997) groups diasporas into five types. They are: victim/ refugees (Jews, Africans, Armenians, Palestinians), imperial/ colonial (British, Russian), labor/service (Chinese, Italians), trade/business/ professional (Indian, Lebanese), and cultural/postmodern/ hybrid (Caribbean people) diasporas. It appears that groups in the final four categories may engage in an active transnational life, enjoying the resources to maintain contacts with the homeland and traditional language/culture. The Tamils belong to the first category. They have left Sri Lanka largely in response to the current fighting and human rights violations. They don't have the possibility of traveling often to their homeland. Many of them don't have an interest in going back home even if conditions improve. It appears that there is greater moti-vation for such groups to reconcile themselves to their new life in the West and adopt the languages dominant in the new context for their purposes.

In fact, the unusual case of Tamils compels us to reconsider how diasporas are defined. Cohen (1997) comes up with nine characteris-tics all diaspora communities should display: (1) dispersal from an original homeland, often traumatically; (2) alternatively, the expan-sion from a homeland in search of work, in pursuit of trade, or to fur-ther colonial ambitions; (3) a collective memory and myth about the homeland; (4) an idealization of the supposed ancestral home; (5) a return movement; (6) a strong ethnic group consciousness sustained over a long period of time; (7) a troubled relationship with host societies; (8) a sense of solidarity with co-ethnic members in other countries; and (9) the possibility of a distinctive creative, enriching life in tolerant host countries (180). In this list, Tamils may be considered to display only conditions 1 and 9.[4] But in the context of the frame-work I have developed above, it is clear that Cohen's definition is informed by traditional assumptions of community, identity, and nationhood. It is based on essences, grounded in territorialist features, and informed by bounded modes of wholeness. A postmodern orien-tation to diaspora will be defined according to a different set of assumptions. A definition of diaspora based on in-betweenness, fluidity, and open-endedness is now needed. Such diasporas will be detached from their territorial moorings, evolving, incomplete, and proliferat-ing. Such is the definition of diaspora that Stuart Hall (1990) adopts: "The diaspora experience as I intend it here is defined not by essence or purity, but by the recognition of a necessary heterogeneity and diversity; by a conception of identity which lives with and through, not despite difference; by hybridity. Diaspora identities are those which are constantly producing and reproducing themselves anew, through transformation and difference" (235). Hall's own black diaspora is a good example of such a postmodern community. Detached from any mythic notions of returning to Africa, black communities in the Caribbean islands, South America, and United States have defined their new identities in a formerly colonizing language. As to be expected, Cohen is a bit uneasy about accommodating such communities into his notion of a diaspora.

Perhaps I have to change my attitude that the emerging picture of the Tamil diaspora is disappointing as it lacks the dazzling forms of cultural hybridity and linguistic diversity that we find in other transnational communities. It is possible that the sociolinguistics of such newer "victim" diasporas will be more interesting because it is more subtle.

NOTES

* I gratefully acknowledge the funding from PSC-CUNY research awards # 27, 28, and 29 to collect data relevant to this research. An NEH summer stipend in 1998 enabled me to transcribe and analyze parts of this data.

1. The statistical data presented here from the questionnaires relate to the information gathered from Lancaster, California. The survey data from the other locations are still being analyzed. My comments on other locations (London and Toronto) are based on the data from field work notes and interviews.
2. See also Wenger 1998.
3. Recently, Chase Hensel (1996) has adopted a similar construct to explain how Yup'ik Eskimos of Southwestern Alaska construct membership according to those who can participate in their activities of hunting, fishing, and processing.
4. Condition 7 may also be fulfilled in certain locations. But Tamils are still not conspicuous enough to attract focused forms of discrimination—at least according to the community members I spoke to. Condition 2 is fused with 1, as many come to the West expecting better social prospects, besides fleeing the ethnic violence.

REFERENCES

Appadurai, Arjun. 1996. *Modernity at large: Cultural Dimensions of Globalization.* Minneapolis: University of Minnesota Press.

Bhabha, Homi K. 1994. *The Location of Culture.* New York: Routledge.

Boyd, Sally and Sirkku Latomaa. 1999. Fishman's theory of diglossia and bilingualism in the light of language maintenance and shift in the Nordic

Region. In Guus Extra and Ludo Verhoeven, ed., *Bilingualism and Migration*, 303–324. Berlin: Mouton.

Canagarajah, A. Suresh. 1995. The political-economy of code choice in a revolutionary society: Tamil/English bilingualism in Jaffna. *Language in Society* 24, 2: 187–212.

Canagarajah, A. Suresh. 1999. *Resisting Linguistic Imperialism in English Teaching*. Oxford: Oxford University Press.

Cohen, Robin. 1997. *Global Diasporas*. Seattle, WA: University of Washington Press.

Hall, Stuart. 1990. Cultural identity and diaspora. In *Identity, Community, Culture, Difference*, J. Rutherford (ed.). London: Lawrence & Wishart. 222–237.

Hall, Stuart. 1997. The local and the global: Globalization and ethnicity. In *Culture, Globalization, and the World System*, A.D. King (ed.). Minneapolis, MN: University of Minnesota Press. 19–40.

Hensel, Chase. 1996. *Telling Our Selves: Ethnicity and Discourse in Southwestern Alaska*. New York: Oxford University Press.

Lave, Jean and Etienne Wenger. 1991. *Situated learning: Legitimate peripheral participation*. Cambridge: Cambridge University Press.

Mignolo, Walter D. 2000. *Local Histories/Global Designs: Coloniality, Subaltern Knowledges, and Border Thinking*. Princeton: Princeton University Press.

Papastergiadis, Nikos. 1998. *Dialogues in the Diasporas*. London: Rivers Oram.

Papastergiadis, Nikos. 2000. *The Turbulence of Migration*. Cambridge, UK: Polity Press.

Prior, Paul. 1998. *Writing/Disciplinarity: A Sociohistoric Account of Literate Activity in the Academy*. Mahwah, NJ: Lawrence Erlbaum.

Tosi, Arturo. 1999. The notion of "community" in language maintenance. In *Bilingualism and Migration*, Guus Extra and Ludo Verhoeven (ed.). Berlin: Mouton. 325–344.

Wenger, Etienne. 1998. *Communities of Practice*. Cambridge: Cambridge University Press.

PART 4

CONNECTIONS

9

TEACHING ENGLISH AMONG LINGUISTICALLY DIVERSE STUDENTS

John Baugh

INTRODUCTION

Educators are at a considerable disadvantage when it comes to teaching students from linguistically diverse backgrounds. Policies that have been formulated by federal, state, and local school officials rarely meet or match the educational needs of students within schools. This is especially the case in schools that serve highly diverse populations.

I begin with some personal reflections, because I attended inner-city public schools in Philadelphia and Los Angeles where competing linguistic pressures were evident on a daily basis. That personal foundation sets the stage for considering some contemporary linguistic controversies pertaining to many African American students; namely those who speak African American Vernacular English (AAVE), and traditional language minority students for whom English is not a native language (ENN).

Although a host of educational regulations exist that are intended to support the academic welfare of linguistically diverse students, we will see that they are inadequate to the task and severely restrict educational choices among students who lack proficiency in Standard English. Some of these barriers are outlined in considerable detail, thereby exposing the potential for educational malpractice.

In an attempt to provide an alternative approach, I review the linguistic heritage of the United States, and corresponding linguistic diversity and/or linguistic homogeneity within American classrooms throughout the country. These linguistic facts are then compared and contrasted with existing educational policies, which further expose their woeful limitations.

Various solutions are considered along with evidence of successful educational programs at an inner-city high school that serves minority students from a broad range of cultural and linguistic backgrounds. Students who attend this school are described at considerable length by Alim (2003), and his findings indicate that students maintain strong cultural and linguistic loyalties to their heritage language, but they do so with an explicit goal to excel academically. The existence of such students, who are the beneficiaries of creative and innovative educational programs, call existing regulations for low-income and language minority students into question.

SOME PERSONAL REFLECTIONS

My formal education began in Philadelphia. The vast majority of my classmates were African American, but not exclusively so. There were some working class white students, the rare Asian American, and no Latinos/as. All of my teachers were white, and few of them were familiar with African American culture or language. Some of my earliest memories of schooling were of hard-nosed teachers who were determined to help me and my fellow classmates "speak properly," and, by extension, to become literate. I also recall feelings of embarrassment on occasions when my papers were exposed to public ridicule as examples of "what not to do." I should hasten to add that I completed all of my assignments to the best of my inexperienced ability.

Our third grade teacher had a volatile temper and my most vivid memory of schooling at that time was one of trying to avoid her wrath. Regrettably, I was only partially successful, and some of these episodes centered on language. In one instance, having learned the silent "e" rule, I mistakenly spelled "climb" as "clime," thinking that the silent "e" would correspond to the long vowel sound, similar to "like," "nice" or "file." Obviously, I was wrong, but I recall my sense of incredulity when my paper was marked wrong with a spelling of "climb." I knew full well that "b" and silent "e" sounded quite different, as in "club" or "clue."

Although I was genuinely perplexed, trying to make linguistic sense of facts that struck me as odd, or worse, unfair, the teacher became agitated and increasingly angry as I asked for clarification and reconciliation of the rules that I thought I had previously understood. It was readily apparent that the teacher was frustrated, and my persistent questions only added fuel to an incendiary situation. I never intended to evoke anger; I merely wanted clarification, especially since I knew that the "b" sound never (in my experience) mirrored that of silent "e."

The illustration at hand has nothing whatsoever to do with AAVE; speakers of Standard English have all encountered the special circumstances surrounding the mismatch between the spelling of "climb" and its pronunciation. However, this episode illustrates the atmosphere of linguistic intimidation that I and my fellow AAVE speaking peers faced at the hands of a teacher who was ill equipped to meet our educational needs.

The following year my family moved to central Los Angeles, and for the first time I was exposed to many classmates who were learning English as a second language. Several were native speakers of Chinese, Japanese, and Spanish, and others had learned English natively, albeit as nonstandard dialects owing to their family heritage.

In Los Angeles about one-third to one-quarter of my classmates were African American, and we felt that we had clear linguistic advantages over students who were struggling to learn English. Teachers at this school, which included my first and only African American teacher, were less inclined to admonish the English of African Americans, that is, in comparison to our classmates who were ENN.

Upon reflection, with all of the advantages accrued with hindsight, I believe that most of my elementary school classmates in Philadelphia and Los Angeles faced considerable language barriers that impeded our academic progress. At that time, however, I was often made to feel—as were my peers—that we were willfully misusing English—to the detriment of the language, if not ourselves.

CONTEMPORARY LINGUISTIC CONTROVERSIES

I entered school in 1955, a year after the landmark *Brown v. Board* Supreme Court ruling. At that time educators were not directly concerned with matters of linguistic diversity, because racial divisions were overt under programs where de facto or de jure segregation existed.

More recent educational controversies have centered on Ebonics, growing from a notorious educational resolution that was passed by the Oakland, California, school board declaring that Ebonics was not English, and that it was the native language of 28,000 African American students within that school district. Elsewhere I, and others, have surveyed that controversy at considerable length (Adger et al. 1999; Baugh 2000; Perry and Delpit 1998; Rickford and Rickford 2000; Smitherman 2000).

Beyond Ebonics, a great deal of energy has been devoted to micromanaging bilingual education programs, or, more precisely, some have attempted to dismantle bilingual education by severely restricting the educational options for ENN students (see August and Hakuta 1997; Baugh 1998; Crawford 1992; Zentella 1997).

Throughout the history of the United States there have been various efforts to entice and/or coerce residents to adopt English as the common national language. Many correctly point to the communicative virtue of a common national language as a means to enhance public discourse, cohesion, education, and commerce; however, as a nation of immigrants, it is not readily evident how best to achieve this linguistic commonality.

Perhaps the most visible recent effort to control and restrict educational options among students for whom English is not native has been spawned by Mr. Ron Unz, who is a software developer from Palo Alto, California. Mr. Unz is independently wealthy and has taken considerable umbrage with bilingual education, so much so that he has championed various voter propositions to dismantle bilingual education in various states throughout the nation.

Without the benefit of any formal training in linguistics or education, Mr. Unz and some of his like-minded compatriots were openly critical of bilingual education, claiming that such programs systematically retarded the academic development of English language learners, and should be abolished in favor of immersion programs that would place language minority students in classrooms where English prevails as the unequivocal dominant language.

As Mr. Unz gained publicity for his cause, advocating "English for the children," many bilingual educators countered his efforts by attempting to defend programs and policies that were intended to advance the status quo. Mr. Unz pointed to patterns of academic failure that have afflicted the vast majority of language minority students. As far as he was concerned, bilingual education was largely to blame for this dismal academic performance of children who were English language learners.

He has waged a tireless campaign against bilingual education, and in the process he has, perhaps inadvertently, vilified many dedicated bilingual education teachers by portraying them as self-serving ideologues who are more concerned with their paycheck than they are for the welfare of their ENN students.

Taken together, the Ebonics controversy and Mr. Unz's efforts target students who for a variety of reasons have not developed linguistic competence in Standard English. Of greatest concern to this author is the fact that these are the very students who are most in need of linguistic enrichment to improve their academic prospects. In the case of Ebonics, Oakland's efforts to call attention to the educational plight of African American students who lack Standard English proficiency were abruptly suspended in the wake of public ridicule across racial lines. Other school districts have been loath to follow in Oakland's footsteps, thereby leaving the linguistic plight of the vast majority of African American students in abeyance.

Unlike African American students who do not speak Standard English, ENN students are supported by a combination of federal, state, and local efforts across America, but the jury is still out regarding several legal, economic, and policy matters that pertain to English language learners who attend public schools. If Mr. Unz has his way all such students will rely exclusively on rapid transition into mainstream classes through English immersion only.

This "one size fits all" approach overtly restricts the educational options available to ENN students, and it fails to recognize many of the other barriers that impede their academic progress. Because Mr. Unz has devoted his efforts to ballot initiatives, the vast majority of voters who will proclaim the educational fate of language minority students will never be subject to these provisions themselves.

These issues are hotly debated in their respective home communities; African Americans have strong and diverse opinions about Ebonics, and many U.S. residents and citizens who are native speakers of languages other than English are also divided as to the best ways to advance academic success among students who share their linguistic and cultural heritage.

RELEVANT EDUCATIONAL POLICIES

At present, major educational reforms have been proposed under President Bush's initiatives, which are more popularly known as "No Child Left Behind" (NCLB). Prior to the implementation of NCLB, individual schools or school districts applied directly to the federal government for financial support for language minority students, and therein lay the peril that Oakland faced during the Ebonics controversy. Stated in other terms, should African American students who lack Standard English proficiency be considered "language minority students?" Former secretary of education, Richard Riley, emphatically said "No" to the preceding question. He did so indirectly, however, by stating that "no bilingual education funding should be used for speakers of Black English."

His use of the term "Black English" rather than "Ebonics" makes clear that he does not consider African American students to speak a language other than English. Of greater policy significance, only those students who spoke languages other than English were eligible for Title VII funding, that is, the very funding that was requested of the federal government through direct applications from schools and school districts throughout the country.

Had Oakland educators been successful in their attempt to procure bilingual education funding for their African American students, it is highly likely that every major school district with a substantial population of African American students would follow their lead. Indeed, it was—in part—due to a desire to lead the nation toward a more inclusive and comprehensive policy regarding the education of language minority students that Oakland educators put forward their controversial resolution in the first place.

By striking contrast, traditional language minority students, that is, those who learned a language other than English as their mother tongue, were covered under Title VII. But because of the requirement that schools or school districts submit formal application for this funding, access to bilingual education was not entitled, and the ensuing hit-or-miss access to bilingual education proved to be problematic.

NCLB attempts to resolve some of these inconsistencies, through revisions in Title III, which will now direct bilingual education funding to the states, which in turn will distribute them to schools that serve populations of ENN students. NCLB is far more comprehensive in its scope than implied by my remarks here, which are devoted almost exclusively to language education; readers who wish to know more about this topic can do so by consulting directly with the United States Department of Education, either by mail or through their Internet website.

During the past thirty years, and perhaps longer, federal, state, and local policies pertaining to students for whom Standard English is not native (SENN) have been a moving target, and that target can move, yet again, in different directions from state to state. There are advantages and disadvantages to this reality. The fact that individual states have the legal flexibility to create innovative educational policies for students who reside within their boundaries is, in my opinion, a good thing. However, there is no guarantee that states will devote sufficient funds or attention to their neediest students, thereby revealing a weakness. The federal government does not establish national standards beneath which no state shall fall. Thus, states that lower the bar on public education are free to do so, and their citizens suffer as a consequence. Though it may be controversial to consider, the federal government has a responsibility to set minimum educational standards beneath which no school should ever fall. This may seem counterintuitive, as NCLB and so many other educational ventures seek to establish high standards. The point at hand is fairly straightforward; unless cognizant authorities clearly spell out the minimum educational standard, then efforts—no matter how well intended—to elevate educational standards may be misleading because of the absence of an intellectual safety net to truly ensure that no child is left behind.

A MODEL OF RESOURCE ALLOCATION

To further illustrate this point, I consider two educational resources that are instrumental to every school in the nation, if not the world; namely, the allocation of time and money. Figure 9.1 illustrates variability along these lines in an effort to illustrate how educators, legislators, or school board members might go about considering these allocations. A brief thought experiment further illustrates the point. Consider, if you will, your own impression of more-than-adequate to less-than-adequate allocations of time and money for educational purposes. You may reflect upon your own education, perhaps that of your children, or any students that come to mind.

Elsewhere (Baugh 1999) I have argued that schools illustrated by A3, C1, B3, C2, and C3 should be thought of as unacceptable for educating students, yet many schools operate within these constraints. Indeed, how egregious must a case of inadequate funding or time allocations be before it constitutes educational malpractice?

Adequate time <------------> Inadequate time

(+) F U N D S (-)	A1	B1	C1
	A2	B2	C2
	A3	B3	C3

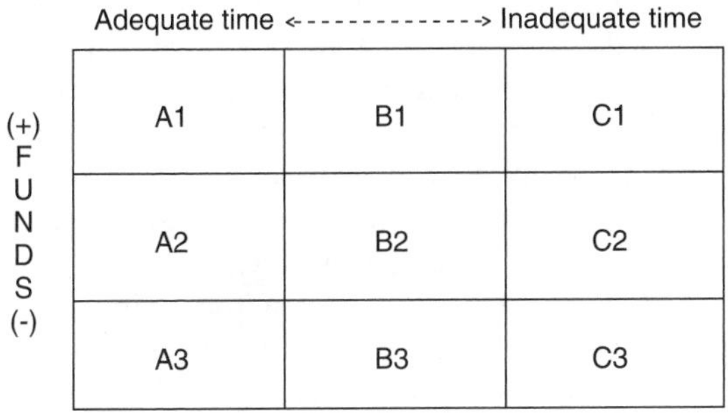

Figure 9.1   A model of resource allocation: relevance to educational malpractice.

Under the proposed model, we would all want our children to attend schools classified by A1, where more than enough time and funding is available, and, at the opposite end of the spectrum, schools that fall within the C3 category run the risk of fomenting miseducation, often through no fault of qualified teachers who are simply helpless to overcome external limitations that are beyond their power to control.

MATTERS OF RELEVANCE TO LINGUISTIC DIVERSITY IN THE UNITED STATES

The preceding illustration is not merely offered to engage your intellectual curiosity, it also corresponds to another sociolinguistic reality; namely, students who come from affluent homes where Standard English is the norm are far more likely to attend A1 schools. Despite glaring educational needs, funding allocations for schools are such that students who do not reside in affluent homes where Standard English is the norm are far more likely to attend C3 schools.

When we combine sociolinguistic considerations with the existing allocation of educational resources we find a disturbing trend; students who are most in need of help are very likely to be attending schools where that help is least likely to be forthcoming. Indeed, it is because of this reality that educational initiatives such as NCLB have been proposed. Time will tell if the effort has been sufficient to meet the massive challenges affecting all students who attend America's schools, that is, regardless of their linguistic background.

Although some efforts and advances have been made to increase teacher professionalism, such as through the National Board for Professional Teaching, these efforts are not yet systematic, nor are all teacher preparation programs equally well equipped to provide their graduates with the necessary tools and techniques that will allow them to help students from diverse backgrounds.

At an absolute minimum, teachers across all subject areas should be trained in procedures that will increase their awareness and effectiveness with all students. Anything less in the training of America's teachers will simply perpetuate a system that is biased in favor of affluent students who already have substantial linguistic advantages derived from residing in homes where the dominant linguistic norms prevail.

OVERCOMING THE LINGUISTIC PARADOX
OF AMERICAN EDUCATION

The educational paradox to which I refer has been described above. Children who lack proficiency in Standard English are very likely to attend schools that are least well equipped to meet their needs. Moreover, if outsiders to those schools and communities can mandate how to educate language minority children, then poor families will experience even fewer educational opportunities than they do now.

The solution to this educational paradox lies within a strategic allocation of human and fiscal resources within a local community, if not within the community of the local school itself. I have yet to visit an American school where parents did not have high aspirations for their children who attended that school. I also have observed that some parents are far more active than others when it comes to demanding services for their children, and these demands often correlate with the wealth and educational experience of the parent(s).

Parents who themselves have met with considerable success in school are more likely to be direct and vocal advocates for their children, whereas parents who have not met with success, or who are unfamiliar with the nature of American schools, are far less likely to be active educational advocates for their children, and therein lies both a concern and an opportunity.

A trilateral partnership must exist for education to work effectively: educators, parents, or guardians, and the students themselves must have avenues of communication and mutual devotion to the academic well-being of students. If any leg of this three-legged stool is weak, then the entire enterprise runs the risk of collapse.

Educators and school leaders who seek to educate all students, regardless of linguistic background, will need to make the necessary arrangements to communicate effectively with parents and students for whom Standard English is not a norm. In schools and school districts with limited funding this effort will require the strategic utilization of human and modest fiscal resources. To what extent, if any, are efforts being made to inform parents and students who lack Standard English proficiency? If the answer is "little" to "none," then it should come as no surprise when students' academic performance remains low.

Parents who do not take full advantage of opportunities to communicate with teachers and other educators affiliated with their local schools also run the risk of perpetuating lower academic achievement. For example, if schools are making direct and culturally appropriate efforts to reach parents and guardians, and they meet with an inadequate response, then it is the students who suffer.

Of ultimate concern is the student; she or he must put forth their best effort, and do so with consistency if they hope to succeed in school. The best communication between educators and parents will amount to very little if students do not do their part through devoted hard work. While these suggestions are certainly not new, and embody well-worn platitudes, they are seldom applied with equal vigor to students from diverse language backgrounds, and it is on this score that I hope these observations may have a greater impact.

Government at the federal, state, and municipal levels can support local schools in ways that allow teachers, parents, and students to tailor their local curriculum to meet high academic standards, and to do so in ways that best serve the constituency of students that populate any given school. Mandates that treat all ENN or SENN students as if they are cut from the same cloth, or which fail to recognize that they often attend schools with limited resources, are unlikely to make the necessary difference that will improve their educational prospects.

REFERENCES

Adger, Carolyn Temple, Donna Christian, and Orlando Taylor (eds.). 1999. *Making the Connection: Language and Academic Achievement among African American Students: Proceedings of a Conference of the Coalition on Language Diversity in Education.* Washington, DC: Center for Applied Linguistics; McHenry, IL: Delta Systems Co.

Alim, H. Samy. 2005. *You Know My Steez: an Ethnographic and Sociolinguistic Study of Style Shifting.* Publication of the American Dialect Society, #89. Durham, NC: Duke University Press.

August, Diane and Kenji Hakuta (eds.). 1997. Improving schooling for language-minority children: A research agenda; Committee on Developing a Research Agenda on the Education of Limited-English-Proficient and Bilingual Students, Board on Children, Youth, and Families. Commission on Behavioral and Social Sciences and Education, National Research Council, Institute of Medicine. Washington, DC: National Academy Press.

Baugh, John. 1998. Linguistics, education, and the law: Educational reform for African-American language minority students. In *African American English: Structure, History, and Use*, Salikoko Mufwene, John Rickford, Guy Bailey, and John Baugh (eds.). London: Routledge Press. 282–301.

Baugh, John. 1999. *Out of the Mouths of Slaves: African American Language and Educational Malpractice.* Austin: University of Texas Press.

Baugh, John. 2000. *Beyond Ebonics: Linguistic Pride and Racial Prejudice.* New York: Oxford University Press.

Crawford, James. 1992. *Hold Your Tongue: Bilingualism and the Politics of English Only.* Reading, MA: Addison Wesley.

Perry, Theresa and Lisa Delpit. 1998. *The Real Ebonics Debate: Power, Language, and the Education of African American Children.* Boston: Beacon Press.

Rickford, John Russell and Russell John Rickford. 2000. *Spoken Soul: The Story of Black English.* New York: John Wiley & Sons, Inc.

Smitherman, Geneva. 2000. *Talkin' that Talk: Language, Culture and Education in African America.* London: Routledge Press.

Zentella, Ana Celia. 1997. *Growing up Bilingual: Puerto Rican Children in New York.* Malden, MA: Blackwell.

10

PLAYING WITH RACE IN TRANSNATIONAL SPACE: RETHINKING MESTIZAJE

Marcia Farr

Los Cárabes vinieron de España, los primeros como detectives . . . El rey de España los mandó a buscar los restos de un sacerdote o un . . . fraile que había muerto aquí. Le dijeron a dos personas Cárabes, "Tú vas . . . a esa parte, aquí está el mapa, consigues dónde enterraron esos—el cuerpo de aquella persona y me traes los huesos."

Tenían que investigar 'ónde había sido, 'ónde lo . . . posiblemente lo haigan matado o se murió, pero allí lo enterraron. Y el rey . . . o la reina quería los huesos de ese fraile allá. Duraron parece que nueve o once años. Pero lo llevaron, uno, y otro se quedó. Y el que lo llevó allá duró, cuando pudo . . . regresó, pero regresó a Michoacán. Le gustó aquí la tierra, la . . . las güares {laughs}.

Y los mandaron a ellos porque eran hombres muy vivos. Ya tenían misiones cumplidas en ese ramo . . . Salieron de España, llegaron a méxico . . . se quedaron en Michoacán . . . y tuvieron familia con la Malinche si tú quieres {laughs} y y así se fue el apelativo . . . siguiendo. Y de esa manera se extendieron los Cárabes. {Chuckle}. D' ese es la . . . descendencia de los Cárabes.

The Cárabes came from Spain, the first ones as detectives. The king of Spain ordered them to look for the remains of a priest or a . . . friar who had died here. They told two Cárabes men, "You go . . . to that part, here is the map, find out where they buried those—the body of that person and bring me the bones."

They had to investigate where it had been, where it, possibly they had killed him or he died, but there they buried him. And the king . . . or the queen wanted the bones of that friar there. It seems they lasted nine or eleven years. But they took [the body back], one [of them], and the

other stayed. And the one who took it back, stayed there, [and] when
he could, he returned, but he returned to Michoacán. He liked the land
here, the, the *güares* [indigenous[1] women]. {laughs}
And they sent them because they were very quick-witted men. They
already had completed missions in that line [of work] . . . They left
Spain . . . they arrived in Mexico . . . they stayed in Michoacán . . . and
had family with *La Malinche*, if you want {laughs}, and, and so the sur-
name continued. And in that way the Cárabes spread out {chuckle}.
That is the . . . ancestry of the Cárabes.

These words were spoken to me by a man at the kitchen table in his
house in the *rancho*. He was the first from the *rancho* to travel to
Chicago to work, coming as a contract worker in 1964, during the last
year of the U.S. Bracero Program. He is now retired and living back
in the *rancho* where his forebears have lived for centuries. He, like
many other adults in this social network of families, carries on his life
both in the *rancho*, where he now spends most of his time, and in
Chicago, where (at this writing) four of his six children live, most of
them raising families themselves. (One daughter lives with her husband and son in California, and another daughter, with her husband
and two children, recently returned to live once again in the *rancho*,
after over twelve years in Chicago.) This man and / or his wife
frequently visit Chicago, sometimes staying for months at a time (e.g.,
around the birth of a child), and their children's families in the United
States regularly return to the *rancho*, for several weeks' vacation, or
even for several months to work on special projects (constructing their
own houses, helping in the family avocado orchards at crucial times of
the year, attending weddings and other fiestas). These visits, of course,
are constrained by work and school schedules in Chicago.
Construction workers, for example, who sometimes are laid off in
Chicago's harsh winter weather, have the flexibility, which comes with
no paycheck, to extend their stays in Mexico. Those women and men
who work in factories usually don't have such flexibility, unless they
too are laid off, or they quit their jobs, intending to find new ones
when they return to Chicago. Sometimes, however, relatives work
temporarily in the place of those who go to Mexico for an extended
visit, especially when employers want to retain valued employees.
Children enrolled in Chicago public (or sometimes Catholic
parochial) schools generally are restricted to Christmas, Easter, and
summer vacations for their returns to the *rancho*. Preschool children,
however, unconstrained as yet by school schedules, are sometimes
sent to be with their grandparents for extended periods. Moreover,

many children even into the third generation still are socialized partly in the *rancho*. This pattern, of course, varies, as some children go to school entirely in Chicago, and grow up and begin to work in Chicago, making acquaintances and friends outside the social network in the process. Nevertheless, there is a tie to the *rancho* that extends into the third generation for most families in the network.

This tie is not surprising, considering the fact that this family can be traced back at least three centuries in this micro-region in northwest Michoacán. The story related above is the oral tradition that traces the family's origins to Spain, an oral tradition that was told to this man, the eldest brother in his natal family, by his father, who presumably learned it from his own father. According to González Méndez and Ortiz Ybarra (1980), some of this man's male ancestors migrated from an area near Cotija, a town in the western part of this micro-region, to found a *rancho* just up the road from this one. Cotija was, in turn, the destination of *ranchero* families from Los Altos of Jalisco (Cochet 1991), which Barragán (1997) terms the distant cradle of *ranchero* society. Thus the *ranchero* identity evident in these families can be traced to their own family histories. Many *ranchero* families from this area, in fact, trace their ancestry back to Spain (and one prosperous family in the *rancho*, with professional members in Guadalajara, has a Spanish coat of arms on the wall of their architect-designed house), although most people readily acknowledge that their ancestors (and those only a few generations back, after the Revolution of 1910–1920) "mixed the blood" with indigenous Mexicans. In a conversation among several women the morning after we had all spent the night in the "female" bedroom of her home,[2] a young woman whom I know well remarked that *mestizaje* in the *rancho* is reputed to be relatively recent, having occurred primarily since the Revolution. Others outside of the *rancho* also have indicated that this *rancho* was known for being populated by whites.[3] Although most people in the *rancho* acknowledge a partially indigenous heritage and thus would be categorized as *mestizo* (racially mixed), many individuals and even entire families in this *rancho* are perceived as "white" in the United States, until they speak Spanish or Spanish-accented English. That is, many people have blue or green eyes, blond or light brown hair, and light skin (with freckles) that turns red, not brown, in the sun. Others look more evidently *mestizo*, with tan skin and some features (e.g., turned up rather than straight and narrow noses) that are characteristically indigenous in this region.

Archival references date this man's family name to the eighteenth century in this micro-region, and oral history interviews have linked his father through kinship to a wealthy *ranchero* (with the same name) in the early part of this century.[4] Photos show this family forebear to have light eyes, skin, and hair. Several different people, both within and outside of this family, have traced recent *mestizaje* (race mixture) in this family to a grandmother and a great grandfather of the man who told the story quoted above; one of these forebears had both Spanish and Indian ancestry and the other had Spanish, Indian, and African ancestry.[5]

This "origins" story, however, illustrates an intensely felt non-Indian (and non-African) identity. During my fieldwork both in Chicago and in the *rancho* over a period of ten years, this sense of identity emerged in countless conversations. Thus while easily acknowledging their *mestizo* heritage, the families from this *rancho* identify nonequivocally as nonindigenous, a claim that is supported by the physical appearance of many individuals and families. People here sometimes refer to themselves as *blancos* (whites); for example, one man said to me, in reference to distant relatives from another *rancho*, *Son blancos como nosotros* (They are whites like us). The fact that many of these *rancheros* are light-skinned and light-eyed, even blond (*güero/a*), attests to the presence of Spanish, French, and possibly other Europeans in these parts in the past. French troops, for example, were stationed contiguous to the *rancho* for several years during the French Intervention in Mexico (1862–1867), and, among other French influences on Mexico, the nearby *municipio* (county seat) produces *pan blanco* (white bread), also called *pan de vapor* (steamed bread), that closely resembles what is called French bread in the United States.

Although the Spanish of this network is lightly sprinkled with *Purhépecha* (the indigenous language in northwest Michoacán) words, especially words for various types of soil and place names, such borrowing of vocabulary is not unusual in language-contact situations. Beyond vocabulary items, some individuals can sing particular songs in *Purhépecha*, especially the well-known *Flor de Canela* (Flower of Cinnamon; note the Spanish title), but this knowledge is framed as Other and kept separate, which only confirms a primarily nonindigenous identity. Thus these *rancheros*, like others in pockets all over western Mexico, construct themselves as nonindigenous, even while acknowledging their *mestizaje*. Knight argues that such claims are particularly vocal in contexts (especially "Indian zones" such as northwest Michoacán) in which "lower-class mestizos . . . cleave . . . to their eroding ethnic privilege" as Indians begin to compete with them economically and socially (Knight 1990, 99). Though Knight offers a plausible explanation from a research perspective, when this racial ideology is explored for emic meanings at the local level, it is revealed as more than a simple claim to higher social status. It calls into question the category *mestizo* itself. Although this term is widely used in the research literature on Mexico, it is not a category that is emically derived, at least not for all so-called *mestizo* communities. Rather, it is a term created from outside such communities that is closely tied to colonial racial ideology, and, as such, evokes that ideology when used.

Although "race" has been shown to be a social construct rather than a genetically determined category (AAA Statement on Race 1998; AAPA Statement on Biological Aspects of Race 1996), conventional thinking continues late-nineteenth-century notions of "races" (e.g., in U.S. Census forms),[6] and the terms *mestizaje* (race mixture) and *mestizos* (those who are "mixed" racially) invoke this ideology when used. In this sense, it is difficult to "think outside the language"—that is, as long as we use these terms, we perpetuate the assumption that separate races exist, and that they have "mixed." Research literature on Mexico often specifies whether the site of a particular study is an "Indian" or a "*mestizo*" community, for example. I have rarely, however, heard the word *mestizo* used by the *rancheros* of this study to refer to themselves. When asked about *mestizo* communities, they refer to formerly indigenous communities that have gradually become hispanicized over time, through the "crossing" of Spaniards and Indians. Such distancing from *mestizaje*, even while acknowledging some indigenous heritage themselves, clearly reveals ambiguities and locally perceived differences around racial identity among what are lumped together and generically referred to as *mestizo* communities. After all, communities can become *mestizo* from a primarily Spanish, as well as Indian, base, and these two types of *mestizo* communities can differ sharply on a variety of linguistic and cultural dimensions. As Guillermo de la Peña (1980) has noted, Mexican villages that are physically quite close to one other often contain populations that contrast sharply as social groups and categories within the larger society, and this is especially true in this part of western Mexico. These complexities of identity can be understood better by briefly reviewing *mestizaje* in Mexican history, to which I turn in the next section.

MESTIZAJE IN MEXICAN HISTORY

Race and ethnicity have a complex history in the New World, in the confrontation of Europeans, and Africans, with Indians. After the

Spanish conquest of Mexico, a hierarchical society based on caste, or
"race," was established, with Spaniards at the top, followed by *castas*
(mixed bloods of various types), then Indians, and then Africans.
Although this caste hierarchy evolved toward a more class-based system, especially during the nineteenth century, colonial racial ideology
endured, and it continues to underlie Mexican society even today
(Lomnitz-Adler 1992). Most studies of Mexico and Mexicans have
assumed that *mestizaje*, or racial mixing, has been so thorough that
the two resulting social categories, the (remaining) indigenous
Indians and *mestizos*, are generally indistinguishable from one another
physically. That is, it is widely assumed that there has been so much
genetic, and cultural, mixing in both groups that one cannot tell who
is Indian and who is *mestizo* by physical characteristics alone. And yet
both cultural and genetic characteristics among the families in this
study emphasize a primarily Spanish heritage, illustrating how much
(unstudied) variation exists within the northwest portion of this state,
let alone in the rest of Mexico, or among Mexicans in the United
States.

Another widespread assumption made in most studies of Mexico
that is problematic in this region is that *mestizo* communities have
been, in Bonfil Batalla's term, "de-Indianized" historically. That is,
they were originally indigenous communities, which, through mixture with Spaniards and/or acculturation to "Spanish" culture, gradually lost their indigenous identities (Frye 1996). On a national level,
state ideology since the Revolution of 1910 has promoted Mexico as
a *mestizo* nation, valorizing, at least officially, Mexico's indigenous
past, but working to incorporate nonacculturated Indians into the
mestizo state (Knight 1990; Lomnitz-Adler 1992). Yet in spite of this
public representation of Mexico as a *mestizo* nation, a more complex
variety of identities has endured at the local level, especially in rural
western Mexico.

Several studies of isolated *rancheros* in western Mexico who identify
strongly with the Spanish side of their heritage (Barragán López 1990;
González and González 1974; Taylor 1933) have revealed this complexity. Spanish orientation in such *ranchos* might be the result of the
increased immigration from Spain, especially from northern provinces
(González and González 1995) at the beginning of the eighteenth century. Oral tradition traces the family under study here to one of these
northern provinces, Asturias.[7] Oral tradition in this region also indicates
that *mestizaje* primarily has occurred in recent generations, since the
Mexican Revolution of 1910–1920, and some families in the *rancho* are
experiencing *mestizaje* now, through the marriage of *güeros* (people

with light skin, sometimes blond and blue-eyed) with *prietos* (people with darker skin, sometimes described as swarthy) or *morenos* (people with tan or brown skin). Whatever the historical trajectory of individual *ranchos* and *ranchero* families, however, studies of *ranchero* groups in this region document significant differences from other studies of so-called *mestizo* societies, especially in terms of self-perceived identities and individualist ideologies (Farr 2000).

In contrast to the widely accepted representation of Mexico as a *mestizo* nation, Bonfil Batalla (1996) argues that *mestizaje* in Mexico has not been complete, at least in cultural terms. He critiques the official representation of Mexico (calling it "the imaginary Mexico") as the synthesis of two different cultures, that of Spain and that of indigenous, pre-conquest Mexico. He argues instead that two world-views and civilizational bases simply coexist (though they interpenetrate) in modern Mexico. Instead of a true transformation in culture, what has occurred, he claims, is only a transformation in ideology. The official government ideology promotes images of racial and cultural mixture and integration, but in reality Spanish, or more generally Western, culture has only been superimposed upon a Mesoamerican indigenous base that still underlies most of Mexico, and is what he calls *México profundo* (deep Mexico).

From colonial times to the present, then (and even Bonfil Batalla doesn't question this), a dichotomy has dominated perceptions of Mexico: urban/Spanish/elite versus rural/Indian/peasant. Since much of Mexico's population is rural, Bonfil Batalla claims the predominance of a Mesoamerican (rural) base in Mexican society. According to Bonfil Batalla, although the urban elite disdain what is rural/Indian/peasant, even they have, over time, appropriated some Mesoamerican cultural traits.[8] He further argues that *mestizos* who claim to be non-Indian, especially those who are rural peasants, actually have only been "de-Indianized" superficially, by having had a Western ideology imposed on a basically Mesoamerican civilization. The process of "de-Indianization," according to Bonfil Batalla, is "the loss of these groups original collective identity as a result of the process of colonial domination" (Bonfil Batalla 1996, xviii).

Although Bonfil Batalla in some respects accurately describes the Mexico I have experienced (a "first world" country on top of a "third world" country), he unfortunately generalizes all rural *campesinos* (peasants) as basically "Indian."

In stark contrast to both the imaginary Mexico and *Mexico profundo*, the *rancheros* in this study represent yet another alternative. That is, they, like others in western Mexico, emphasize, and many physically reflect, their Spanish and/or other European heritage. Yet most anthropological research on this region to date has focused either on the indigenous *Purhépecha* (e.g., Friedrich 1977, 1986) or on *mestizo* groups that are presumed to be "de-Indianized" *campesinos* (e.g., Dinerman 1983; Foster 1967/1979/1988). This may reflect a preoccupation with what is presumed to be more "authentic," or a desire to study, and identify with, the most politically and economically oppressed groups, that is, the indigenous. Field (1998), for example, argues that *mestizos* have been understudied and not well understood because they are seen as lacking in cultural authenticity. *Mestizos*, then, either are considered uninteresting because they are not Indians, or they are interesting only because they have an indigenous past. Either way, of course, there is an Othering of Indian-ness, even when positively valorized (and romanticized), that generally ignores communities that are not indigenous or at least "de-Indianized."

In spite of this dominant trend, however, a few studies have documented relatively isolated *ranchero* communities, primarily in western Mexico, that disrupt the stereotype of rural Mexicans as either Indians or de-Indianized *mestizos*. All of the communities in these studies are "*mestizo*," at least to some extent, but they could not be accurately described as "de-Indianized" ones that have "retained" Indian values. Taylor's (1933) early study, for example, describes the "Spanish Mexican" peasant community of Arandas, Jalisco, located in Los Altos of Jalisco, a region directly northwest of northwestern Michoácan and considered the distant cradle of *la sociedad ranchera* (*ranchero* society). Western Los Altos, including Arandas, was colonized by Spaniards at the end of the sixteenth century, and *ranchos* were created there primarily for pasturing sheep, cattle, and horses. The people of Arandas experienced some *mestizaje* with Indians and Africans (the latter brought to the region as slaves), but the non-Spanish contribution to the mixture was slight. Taylor, tracing archival records from the eighteenth and nineteenth centuries, shows a rapid absorption of *mestizos* and mulattoes by dominant whites, both through marriage and through "irregular liaisons."

The work of González and González (1974/1991), Barragán López (1990, 1997; Barragán López et al., 1994), Cochet (1991), and others similarly describe rural communities that are neither indigenous nor de-Indianized, documenting the Mexican saying, *la güera del rancho* (the white/blond of the *rancho*). The present study continues in this tradition, providing a contemporary ethnography of *rancheros* similar in identity to those studied by Taylor, González and

González, and Barragán López, but it departs from this earlier work by focusing on *rancheros* who are less isolated from regional indigenous communities and, moreover, who live within a transnational context.

THE LOCAL SETTING, MICHOACÁN: *NOSOTROS Y LOS OTROS* (OURSELVES AND THE OTHERS)

The *rancho* is nestled amid rolling hills on the edge of what is called the *Meseta Tarasca* (the Tarascan Tableland, or highland plateau) in northwest Michoacán. Tarascans, or in their own language the *Purhépecha*, in the *meseta* primarily live in villages or towns recognized as indigenous. Many adult women wear distinctive skirts, belts, embroidered blouses, and, most significantly, a particular type of *rebozo*, or large woven shawl, which is black with thin bright blue lengthwise stripes and complexly knotted ends. This distinctive clothing, along with a distinct language, *Purhépecha*, mark an indigenous identity in this part of Mexico, and ethnic boundaries between the *Purhépecha* (referred to in Spanish as *Tarascos*) and *rancheros* are scrupulously maintained here.

The status hierarchy of this region of northwest Michoacán places the indigenous *Purhépecha* at the bottom, *rancheros* in the middle, and the urban elite at the top. *Rancheros* mostly live in rural hamlets and make occasional excursions to nearby cities, although increasingly they live in small towns and cities as well. Although people who index a noticeably *ranchero* identity with their clothing, their dialect of Spanish (with archaic rural usage), or other behavior sometimes suffer disdain in the cities, they show disdain for *catrines*, or "citified" people, as well (Farr 2000). Yet except when doing business (e.g., receiving medical services) in cities that have an urban elite population, *rancheros* can avoid most contacts with those above them in the regional status hierarchy with whom they might feel uncomfortable. Primarily they interact with other *rancheros* or with *indígenas*. When interacting with other *rancheros*, their demeanor and language is relatively egalitarian. In interactions with the indigenous, in contrast, *rancheros* expect, and often receive, deference, at least publicly. Friedrich (1977) notes the extreme hostility toward these "outsiders" on the part of the indigenous *Purhépecha* of this region, which suggests that such public deference may be a form of resistance, a "weapon of the weak" (Scott 1990).

Identities are clearest in their contrast with others; in fact, identities are constructed against these others: "we" are not "them." Within northwest Michoacán, *rancheros* and *Purhépecha* distinguish themselves from each other, sometimes fiercely. As Barth noted, it is "the ethnic *boundary* that defines the group, not the cultural stuff that it encloses" (Barth 1969, 15). Otherwise, over time ethnic groups in interaction, as *rancheros* and *indígenas* have been for centuries in Mexico (Barragán López 1997), would tend to exchange "cultural stuff." In fact, such exchange has occurred here in both directions, including the movement of individual people, and yet the boundary between these two groups has remained distinct. *Rancheros* maintain racial boundaries between themselves and the indigenous in both linguistic and nonlinguistic ways. Language is used to distinguish the indigenous either through their use of *Purhépecha* or through the way they speak (their dialect of) Spanish. Boundaries are also maintained through clothing styles, visiting patterns, and the restriction of *ranchero*—indigenous interaction primarily to the commercial domain (e.g., hiring indigenous field workers or buying goods at Indian markets).

Many instances recorded in my field notes illustrate the racial ideology that values lighter skin with which these *rancheros* construct their identity. Talk that indexes a primarily nonindigenous racial identity, then, is frequent within these families, in both Mexico and Chicago, especially among the older generation. Among the younger, more schooled generation, such talk, at least on some occasions, entails ambivalence and acknowledgment of their own (partial) indigenous heritage, since they are taught in federally supported Mexican schools the nationalist ideology that proclaims pride in Mexico's indigenous heritage. In school they are taught that *todos somos indios* (we are all Indians) and *todos somos iguales* (we are all equal). Yet in their families, comments about the indigenous, whether positive or negative, always make it clear that they are different (and usually of lower status). This ambivalence and ambiguity around racial identity is revealed in the analysis of tape recorded discourse in the last section of this chapter.

The Local Setting, Chicago: Racial Categories

Mexican ethnicity in the United States has historically been structured into a disadvantaged minority position; that is, Mexicans as a group have had a disproportionate share of low-level jobs (Nelson and Tienda 1997). Yet this historical legacy is changing: U.S. Mexicans are now principally located in urban areas that have a wider range of employment opportunities. Massey (1981) cites the declining isolation of the *barrio* and a degree of assimilation into Anglo society. For example, there is less residential segregation for Mexicans than for Puerto Ricans and others (Massey and Denton 1989, 1993). Nelson and Tienda (1997) predict that "class divisions could become more salient than ethnicity as Chicanos become more integrated into the non-subordinate part of the labor force," though this depends on the process of immigration and the vitality of the economy. Peñalosa (1995) also claims that "caste" is moving to class, as the Mexican-origin population in the United States becomes more and more class stratified.

Omi and Winant (1994) argue that the United States is moving from a "racial dictatorship" to a "racial democracy," albeit slowly, painfully, and unevenly. They distinguish between race, a social and historical construct that fluctuates in meaning, and racism, the use of "essentialist categories of race" to structure domination. Although racism persists, it is, like all hegemonic projects, incompletely dominant, that is, there are "cracks" in it that allow for challenge (Ortner 1996). From an imagined community of whiteness (Basch et al. 1994, 40) that was used to unite various European groups in a new nation against Others, then, we are moving toward an imagined community of cultural pluralism (Basch et al. 1994), and no doubt toward newer forms of *mestizaje*, mixtures of what are now considered different ethnic and/or racial groups.

Historically, however, racial categories in the United States developed according to a dichotomy between white and nonwhite, with race perceived as being biologically or genetically based (Denton and Massey 1989; Omi and Winant 1994; Rodriguez and Corder-Guzman 1992). The white category itself, of course, emerged in response to the presence of nonwhites, initially Africans and Native Americans, and then Asians and Hispanics/Latinos (Omi and Winant 1994), and black or African American was defined, both socioculturally and legally, by the presence of any African blood (Denton and Massey 1989; Omi and Winant 1994). As Rodriguez (1997) has pointed out, only whites were included in the imagined community of the United States, and recent research by Flores-González (1999) has shown that this imagined community ("real Americans") is still conceived of as white, even by Mexican and Puerto Rican college students in Chicago. Moreover, a categorical perception of race is still evident in the U.S. census item on race, which proceeds from white, to black, then to the rest of the nonwhite categories (Elias-Olivares and Farr 1991), and it is evident as well in the coding procedures that have been used by the U.S. Census (Denton and Massey 1989). Although the 2000 census allowed Americans to indicate more than one race, and put the item regarding Hispanic background before, rather than after, the race item, racial categories still remain.

When Mexicans, or other Hispanics/Latinos, migrate to the United States, they confront a racial scheme that differs from the one they are familiar with in their countries of origin. In contrast to the categorical view of race in the United States, in Latin American countries, including Mexico, racial descriptors comprise a continuum, from white to black and/or Indian, depending on the predominant lower status population. For example, in Puerto Rico there has been a negligible presence of Indians and a substantial presence of Africans, as well as white Spaniards, whereas in Mexico, Indians outnumber Africans historically (Denton and Massey 1989). In the U.S. scheme, with a persisting white/nonwhite dichotomy, Mexicans have had an ambiguous place, at times categorized as white and at other times categorized as a nonwhite minority, although regional differences have been significant in this regard. Texas, for example, "where Mexican Americans have come closest to being treated like a racial caste" (Skerry 1993, 20), is quite different from other parts of the Southwest, and the entire Southwest is strikingly different from the Midwest. In Chicago, for example, Mexicans are sometimes treated as yet another "white ethnic" population (especially politically), and Mexican, as well as Puerto Rican, politicians have strong ties to the Democratic machine in the city.

Skerry (1993) claims Mexicans are an ambivalent minority, sometimes identifying as white and other times as a racial minority akin to African Americans. He cites census data from 1980 and 1990 to illustrate this split: in 1980, 53.2 percent of Mexican Americans self-identified as White, 1.8 percent as Black, and 45.0 percent as Other Race; in 1990, 50.6 percent self-identified as White, 1.2 percent as Black, and 48.2 percent as Other Race (Skerry 1993, 17). Left out of his account of so-called ambivalent identity practices among Mexicans, however, are the ways in which identity categories are not chosen, but are imposed. More recent research has shown some second-generation Mexicans and other Hispanics/Latinos in Chicago are beginning to view Latino/Hispanic as a racial category in itself, to which they belong (Flores-González 1999), and DeGenova (1998) indicates a similar kind of racializing of the Latino category among adult Mexicans in Chicago. All research indicates that few Hispanics/ Latinos, even those with a partial African heritage, identity themselves as black (Denton and Massey 1989), possibly because they become aware rather quickly of the benefits of being white and the disadvantages of being black in the United States, and perhaps also because they are not African American culturally. Since the more limited African presence in Mexico was forced to blend into the general *mestizo* population (Lomnitz-Adler 1992), and only recently have researchers begun to identify aspects of an African cultural heritage in Mexico, even fewer Mexicans than other Hispanics/Latinos would identify themselves as black in the United States Consequently, for most Mexicans in the United States the black category is irrelevant for them; only the white and other race categories are potentially relevant.

In a study in Chicago that explored reasons for the undercount of Mexicans during census taking, Elias-Olivares and Farr (1991) found the race item the most problematic of all on the 1988 census form used in the study. Virtually all residents who participated in the study objected to the racial categories listed as choices; 34 out of 39 specifically stated that an option should have been included for their race. The majority of residents (21 out of 39), or 54 percent, chose Other Race. Of the remainder, 10 residents chose White; 4 chose Indian; and none chose Black or Negro. Since virtually all participants in the study had lived in the United States for at least five years (and over half had lived in the United States for at least fifteen years), we can assume that they were quite familiar with U.S. racial categories. Like respondents in other studies (Martin et al. 1990; Rodriguez 1997), most participants in this study did not view themselves as either white or black; rather, they self-identified as Mexican, Puerto Rican, and so on, and even (a few) as Hispanic or Latino, after using the Other Race category. Statements such as "We're not here—we don't count!" were common responses to this item. One resident said, *Pos blanco, quiere decir un americano, ¿no? Completamente* a white person (then white, that means an American, no? Completely a white person).

In the present study, both in Mexico and in Chicago people display an awareness of U.S. racial categories, due both to the heavily intertwined history of the United States and Mexico and to the extensive transnational flows of people, goods, and ideas during the last century. For example, I was referred to on some occasions in Mexico as *güera güera* (really white), to distinguish between me and the similarly complected *güera* women in the *rancho*. Other times, in Chicago, people made categorical references to groups in conversations, for example, to *gente mexicana* (Mexican people), *gente güera* (white people), or *los güeros* (the whites). Similarly, the terms used for African Americans, *negro* (black) or the more polite *moreno* (brown), were used both in Mexico and Chicago. A woman who has long lived in Chicago once

noted in a conversation with me about education that there had been
much racism between Mexicans and blacks at Farragut High School in
Chicago, but that (in 1995) things were better, since they had new
directors at the school. Once in the *rancho* during a conversation, a
man who had worked in Chicago for a few years before being
deported compared the Indians in Mexico with lower-class blacks in
the United States in terms of social problems such as a high birth rate
(certainly an ironic comparison, as this man has five children). In spite
of such stereotypic generalizations, however, it is invariably the case
that members of these families, when they meet individual African
Americans, or U.S. Native Americans, treat them as individuals and
even develop friendships with them, sometimes commenting explicitly
on how important it is not to judge people on first appearances.

Sometimes generalizations about groups link them to specific characteristics, often in a way in which *ranchero* values and identities are
affirmed by contrast with others. In two typical conversations in a
kitchen in Chicago, people noted the differences between whites,
blacks, and Mexicans in terms of the ability to do hard work. One
woman recounted the number of white and black women who "leave
the line" (quitting the factory job of painting mottos and other
material onto plastic items such as glasses) after only a few hours or
days, implying that only Mexican women can endure the very hard
work. Similarly, a man on another occasion in the same kitchen
recounted that only Mexican men could endure the hard work on *el
traque* (the railroad track): *Todos son mexicanos, es que los güeros no—
pa' el traque no . . . No pueden con el trabajo, es muy pesado* (all are
Mexican, it's that the whites, no—on the track no . . . They can't
endure the work, it's very hard).

These *rancheros*, then, are very aware of U.S. categories and use
them to a certain extent, especially since they are not entirely different
(in the racial order) from the colonial racial ideology that still persists
in Mexico. Yet the preponderance of genotypic mixtures is much
more evident in Mexico than it is in the United States for two reasons.
First, Europeans have been more predominant demographically in the
United States than in Mexico, and second, when racial mixture occurs
in the United States, subsequent generations have been considered
either white or nonwhite depending on phenotype. Offspring of
black–white unions have been treated as black, for example (Lazarre
1996), yet no doubt many white Americans with some nonwhite
ancestry (e.g., slight Native American or African ancestry that they
may or may not be aware of), are nevertheless treated as white, based
on their phenotype. In Mexico, in contrast, like many other Latin

American countries, one family will have members with a variety of racial characteristics.

Peñalosa, among other researchers, has noted this, as well as other kinds of diversity among people of Mexican descent, and has advised that, with regard to the study of Mexican Americans, researchers should stop "trying to find the 'typical' or 'true,' and seek rather to establish the range of variation" (Peñalosa 1995, 411). He suggests first differentiating among the Mexican American regional subcultures of the Spanish-descent Hispanos of New Mexico/Colorado, the *tejanos* of Texas, and the Chicanos of southern California. The Midwest, especially urban areas such as Chicago, are yet another regional subculture, one where the Mexican presence has been built entirely by immigration (Año Nuevo Kerr 1977) in the context of a predominantly immigrant milieu (Farr 2003; Holli and Jones 1997/1995). In addition to such diversity within the United States, there is diversity among Mexicans in Mexico, including their various identities as *rancheros*, different groups of indigenous Mexicans, and urban Mexicans of all socioeconomic classes.

Given the diversity among Mexicans, especially phenotypic diversity, how are these varying Mexicans perceived in the United States, with its historically dichotomous system and discrete racial categories? Gamio's work in the 1920s showed that white Mexicans were not segregated, especially if they spoke English. In contrast, darker Mexicans were not allowed into segregated facilities (Peñalosa 1995). More recently, Telles and Murguia (1990) showed that income differences among Mexican Americans could be traced to discrimination based on phenotype; that is, light and medium complexioned Mexican Americans had significantly higher incomes than dark-skinned Mexicans. Some members of the social network in this study have had experiences similar to those reported by Gamio in the 1920s; that is, the lighter-complexioned among them have been taken to be white, at least initially, in a variety of contexts. Several different people have joked, for example, about how easy it is for Mexican *güeros* to cross the border without papers. On one occasion when three members of one extended family were crossing into the United States, the father handed the border officials some papers for himself and his daughter-in-law (who did not in fact yet have her own papers), but the son was not even asked for papers, "Because," he said with a broad smile, "I'm *güero*!" Another story recounts the crossing of a young blond woman from the *rancho* who was told to speak a few words of English in front of the officials (the only words she in fact knew at that point), and, as they had hoped, she successfully crossed into the United States without papers. On a less successful occasion, one young man, another *güero*, was caught by the INS right after he had crossed the border, but only after they discovered he did not speak English. His brother, similarly *güero*, had already successfully made it to Chicago without papers. When the brother who made it to Chicago found a construction job with the Chicago Transit Authority (another version of the predominant male employment on *el traque*, or the track), as part of job orientation he was sent into a room with English-speaking whites and blacks, while other Mexicans were sent into a room to see a video.

Resisting Categories

Although Mexicans in the United States are often assumed (on various institutional forms) to be Hispanic or Latino, an ethnic category (with Other as race), the term Hispanic has never been used to my knowledge by the members of the social network in this study. The term Latino, in contrast, has been used, but only rarely and only by those who live or have lived in the United States. As a category it is contested by these Mexicans both implicitly and explicitly. As Oboler (1995) points out, the terms Hispanic/Latino function as a two-edged sword. On one hand, they are a forced category imposed externally by the U.S. government starting in the 1980s; in the 1970 census, for example, Mexicans were coded as white, according to Denton and Massey (1989). On the other hand, however, even though there is resistance to these labels, their use has provided a platform from which various Latino subgroups have been able to organize to combat discrimination (Padilla 1985). Ground-level resistance to the labels has been shown in several studies, which indicate that most immigrants of Latin American descent prefer to identify ethnically as their nationality, that is, as Mexicans, Cubans, Puerto Ricans, and so on (Elias-Olivares and Farr 1991; Oboler 1995), although this varies for second and third (or more) generation Latinos (Flores-González 1999; Oboler 1995). An exception to this may be (some) Mexicans, who find it especially easy, when speaking Spanish, to use the term *mexicano* regardless of generation.

Resistance to the terms Hispanic/Latino is, first of all, due to the way they homogenize a population that is extremely diverse in racial, socioeconomic, national, cultural, and historical terms. For example, these terms lump together first generation immigrants from various Latin American countries with citizens of Mexican descent whose ancestors were in what is now the southwestern United States before the Mexican-American War of 1848, and with citizens of Puerto Rican descent whose ancestors, as a colony, were transferred from Spain to the United States after the Spanish-American War of 1898 (Oboler 1995). Resistance to these terms is strengthened by the awareness among those with U.S. experience of the stereotypic and denigrating connotations of the terms in U.S. media and discourse that conjure up images of crime, gangs, drugs, high welfare use, and illiteracy. In short, the terms Hispanic/Latino not only homogenize a very diverse group of people who may not feel any natural allegiance to each other, but these labels place individuals in a racial hierarchy, with whites at the top, that defines them as nonwhite. The fact that this category includes Europeans of Spanish descent illuminates the ideology upon which the racial hierarchy is based: the assumed superiority of northern, as opposed to southern and eastern, Europeans. Although explicit statements of this assumed superiority that attribute it to genetic grounds are rarer now than a century ago, the ideology that underlies this categorizing persists in government forms and in the general populace, even among those who do not benefit from it.

Members of the social network in the present study contest the U.S. racial hierarchy in various ways. First, they implicitly question the discreteness of the categories with jokes about some of them being taken as white; their very phenotypic diversity essentially undermines the white/nonwhite dichotomy that is conventionally assumed in the United States. Similarly, they deconstruct the white category by explicitly referring to Italian Americans as Latinos (those marriages within the social network that have been exogenous, i.e., not to other Mexicans, let alone to others from the *rancho* or its micro-region, have been to Italian Americans). One man who was born in Kansas but who retired in the *rancho*, whose daughter married an Italian American she met in Chicago, said to me, *Italianos, pues, son latinos!* (Italians, well, they're Latinos!), referring to shared customs such as spicy food, a focus on the family, and Catholicism. Another man in the *rancho* who has never been to the United States, in referring to the Italian American wife of another man from the *rancho* who lives and works near Chicago, told me that this woman is not *pura güera* (pure white) because she is Italian, in spite of the fact that Italians are clearly (now) considered white in the United States. Such comments echo a belief that was in fact articulated historically in the United States to justify preferential treatment first of Anglo Saxons and then, more generally, of northern (non-Irish!) Europeans (Oboler 1995). Both of these men, perhaps imposing the more finely graded Mexican racial hierarchy on U.S. dichotomous categories, undermine the white category by separating out Italians.

A second way in which members of this social network resist U.S. racial categories also involves the use of the term Latino. Several people have contested this homogenizing label (even though occasionally others have used it positively) by distinguishing themselves from other Latinos. One woman was critical of Puerto Ricans, for example, because, in her view, they did not have family values like Mexicans do, since their children leave home at eighteen like *los anglo sajonas* (the Anglo Saxons). Others have criticized Latinos who don't speak Spanish, calling into question the lumping together of non-Spanish-speaking Chicanos with Spanish-speaking *mexicanos*. One man complained of people who "have a nopal [a Mexican cactus used for food] engraved on their forehead" (i.e., look very Mexican, perhaps with indigenous features) but don't speak Spanish! Another was critical of upwardly mobile Cubans: *Esos cubanos, aquí entró uno de de de quién sabe qué—de barredor en la pinche CTA, y 'orita es el mero jefe ya!* (Those Cubans, one entered here as the, the, the, who knows what— the sweeper in the damned CTA, and now he's the boss!).

NEITHER HERE NOR THERE: PLAYING WITH RACE IN TRANSNATIONAL SPACE

The contesting of U.S. racial and ethnic categories of white and Latino by these *rancheros* is paralleled by their own resistance to being racialized as Indians, or as de-Indianized *mestizos*, in Mexico. As noted earlier in this chapter, they disrupt the conflation of race and class (and rurality) in Mexico by continually affirming their non-Indian identity. When they are again categorized in the United States as a nonwhite Other (now as Latino/Hispanic), they also take issue with it. Given their familiarity with these two different racial hierarchies, and their ambiguous places in each, they sometimes play with the categories, enjoying the ambiguity, and perhaps their own dexterity in sliding from category to category, depending on the context. All such play, of course, deeply questions the validity of the categories themselves. The remainder of this chapter is devoted to an analysis of an instance of this verbal play, a tape-recorded excerpt of joking among women in a van traveling from Chicago to Mexico for the Christmas holidays one year.

This excerpt occurred during a longer joking session among women of various ages in the van. Partly to pass the time, and partly for the sheer pleasure of *echando relajo* (joking around; a way of speaking linked to disorder, pleasure, and verbal art; see Farr 1994, 1998, forthcoming), these women were taking turns at being humorous, moving from topic to topic according to various contributions from the group. According to Reyna, quoted in Briggs (1988, 231), an "immense desire to be verbally adequate" is often realized through humor in *mexicano* culture. Although Reyna was referring to *mexicano* culture in the southwestern United States, it is equally true within this social network of western Mexicans. Often people take the floor during *relajo* to tell narratives that pleasurably entertain and simultaneously function to draw the group together. As a Mexican language and cultural practice, *relajo* affirms group identity and solidarity, and it serves to socialize younger listeners into Mexican, and *ranchero*, ways of speaking and being. In this particular instance, the floor is predominantly shared; in other *relajos* women take turns telling humorous narratives. Here laughter is most notable in two places: first, when L uses Polish and then comments on the progress it signifies (9–15), and second, after B and D play around with the ambivalence of Indian identity in 28–34. Following L's suggestion that, being Mexican, they should know the indigenous language of their region of Mexico (*Tarasco* or *Purhépecha*) in 28, both D and B utter comments that are "double-voiced"—on the one hand, Oh, *sí*, they should know Tarascan (29–30), but on the other, the Indians are, to use a phrase common in the *rancho*, *pinche indios* (damn Indians), as Delia says while giggling in 31. They end by acknowledging their partial Indian heritage: first B playfully insists that she herself is Indian, so don't talk that way about them in front of her, and then D agrees, noting that all of them are Indian, really, for you can see the *nopal* (A Mexican cactus) coming out of their foreheads.[9] In the unstated background of this conversation are two nations, Mexico and the United States, specifically Chicago, and the different racial hierarchies bound up in their sense of nationhood. Their equivocal places in these two hierarchies, and the fact that they travel, and are traveling now, between two different nation-states and racial contexts, intensifies the ambiguity with which they joke. In this part of the transcript made in the traveling van, three women (B, D, and L) *echar relajo*. That there is a larger audience, however, is clear from the fact that another woman comments on the entire topic (W) in 35.

D: *Yo no te veo delgada.*	D: I don't see you as slender.	1
B: *Pues no pero—*	B: Well, no, but—	2
L: *Pero ella quiere más—*	L: But she wants more—	3
B: *Estoy como la calidad del tordo al revés.*	B: I am like a bird, but in reverse.	4

D: {Laughs}	D: {Laughs}	5
L: *Ella quiere tener más . . .*	L: She wants to have more . . .	6
B: *Más piernas. Más /?/.*	B: More legs. More /?/.	7
W: *No, un poquito más pompis.*	W: No, a little more rear end.	8
L: *¿Tú sabes qué es dupa?*	L: Do you know what *dupa* is?	9
B: *¿Es qué?*	B: It's what?	10
L: *¿En qué idioma te estoy hablando?*	L: In what language am I speaking to you?	11
B: *No, no sé.*	B: No, I don't know.	12
L: *Polaco.* {laughs}	L: Polish. {laughs}	13
B: *Ay, en polaco es todo /?/.*	B: Oh, in Polish it's all /?/.	14
L: *Fíjate nomás el progreso.* [ironic tone]	L: Just look at the progress. [ironic tone]	15
Women: {Laughter}	Women: {Laughter}	16
B: *Ya de lo que—¿ya pasástes al qué?*	B: Now from that— now you've passed on to what?	17
L: *No, no todo.*	L: No, not really.	18
W: *¿A cómo /?/?*	W: How /?/?	19
B: *¿Cómo se dice en inglés pompi?*	B: How do you say in English *pompi*?	20
D: *Butt.*	D: Butt.	21
B: *Ya de eso ya pasástes a polaco y todo.*	B: Now from that you've passed on to Polish and everything.	22
Para el próximo año ya vas a hablar chino y {laughter} *chan chan chan.*	Next year you're going to speak Chinese and {laughter} chan chan chan.	
D: *Como el novio de V dice "Yo sí se francés, yo sí sé frances" y le hace V bien callada, "Sí pero cuando se l-se le acaba el francés le entra el italiano."* {Laughter}	D: Like V's boyfriend says, "I can speak French, I can speak French," and V says real quiet, "Yes, but when the French finishes, the Italian begins." {Laughter}	23
B: *¿Por qué? ¿De dónde es él?*	B: Why? Where is he from?	24
Young Women: *¿Es mexicano!*	Young Women: He's Mexican!	25
D: *Pero es puras mentiras, no sabe.*	D: But it's just lies, he doesn't know.	26

B: *Mexicano, hasta las cachas.*	B: Mexican, to the hilt!	27
L: *No, el mexicano va saber pero tarasco.*	L: No, the Mexican is going to know Tarascan.	28
D: *Oh sí.*	D: Oh yes.	29
B: *Oh sí.*	B: Oh yes.	30
D: *Pinche indios.* {Giggling}	D: Damn Indians. {Giggling}	31
B: *Ehi, calmada con los indios, yo soy india.*	B: Hey, take it easy on the Indians, I am Indian.	32
D: *Todos nosotros, todos. No me ves el*	D: All of us, all. Don't you see the damned	33
pinche nopal /?/ que me sale una tuna	nopal /?/ that the fruit comes out here	
ahí.{laughs}	{laughs}	
B: *El nopal* {laughing}.	B: The nopal {laughing}.	34
W: *Ay ay ay.*	W: Ay ay ay.	35
B: *Ay, como son tremendas.*	B: Oh, how audacious you all are.	36

SYNOPSIS AND INTERPRETATION

Immediately preceding this excerpt, the women were joking about comparing their own bodies to the idealized ones found in magazines and other public media. One woman (B) then states, at the beginning of this excerpt, how she would like to change her own body—with more legs and rear end. This reminds another woman of the word for rear end in Polish, something she may have learned through contact with Polish immigrants, the next largest non-English-speaking group in Chicago after Mexicans. (Such contacts occur in neighborhoods, in English classes, or sometimes at work.) After using the Polish word *dupa* (rear end), she asks if anyone knows what this means, and, then, when no one does, tells them its meaning. She ironically links her learning of some Polish with progress, in 15, by saying, *Fijate nomas el progreso* (just look at the progress). Progress, in the sense of moving ahead in life, is an explicitly articulated value of *rancheros* (Farr 2000), which they use to contrast themselves with indigenous Mexicans, who are seen as communal and not progressing. Moreover, progress is an

expected benefit of going to Chicago to work. Cognizant of the fact
that they are returning to Mexico from working in Chicago, L makes
explicit what is perhaps in the back of everyone's mind and jokes
about it, since the move to Chicago, though certainly resulting in
material progress for virtually all of these *rancheros*, has not been without pain or difficulty. This disjuncture is, then, a ripe topic for joking,
and the women reward her comment with much supportive laughter.

The topic of speaking other languages then brings to another
young woman's mind the insistent claim of her sister's boyfriend
that he can speak French (28), which she says her sister quietly and
sarcastically called into question by noting, *Sí pero cuando se l—se le
acaba el francés le entra el italiano* (Yes, but when the French fin-
ishes, the Italian begins). This too is rewarded with much laughter,
at someone who is trying to be more than he is. B, not knowing this
young man, asks where he is from. When the others delightedly cho-
rus, "He is Mexican!" (and Mexican, *hasta las cachas* or to the hilt),
L comments that, really, a Mexican (from their region of the country), if s/he is to know another language, should know Tarascan
(*Tarasco* or *Purhépecha*). D affirms this right away (Oh *sí* in 29),
since the official public discourse in Mexico, promoted by the government and federal schools, valorizes the indigenous languages and
heritage of Mexico. B repeats and affirms D's Oh *sí*. Following a
slight pause pregnant with meaning, D then utters a phrase common
in the *rancho*: [those] damn Indians! This phrase calls into question
the nationalistic racial ideology that they had just affirmed with Oh
sí. Moreover, it gives voice to their shared reality of *ranchero* attitudes toward Indians and their shared assumption of nonindigenous
identity. Here the disjuncture is between a national official discourse
that valorizes an Indian heritage and the local-level reality of what it
means to actually be Indian in Mexico. As rural *campesinas* who are
often Indianized by the elite in Mexico, these *rancheras* are well
aware of the disadvantages that this implies, and of their own family
histories that disrupt the widespread imaginary dichotomy between
Spanish/urban/elite and Indian/rural/poor.

D and B's use of the English *Oh* rather than the Spanish *Ay* in their
Oh *sí* responses, though no doubt unconscious, may not be acciden-
tal. This code switch, though minor, indexes the transnational context
in which they live and within which, at that very moment, they are
traveling. Moving regularly between two nation-states, and two racial
schemes, highlights the differences between them and leads to a deepened sense of relativity. The "place" of these women and their

ranchero families in a racial hierarchy depends entirely on context, and even then, who they actually are (and what some of them look like) disrupts the logic underlying the racial order in both the United States and Mexico.

Although B is phenotypically quite "white," with brown wavy hair, light skin, and blue eyes, her own racial identity is multifaceted, depending on context. On this tape from the traveling van, B seems to affirm Mexican nationalistic racial ideology (we are all Indians), but the fact that she plays with this topic suggests an ambivalence about it. For her part, D, in revoicing a common *ranchero* put-down of Indians, clearly questions their automatic affirmation of the government ideology. Yet perhaps she also is critiquing the *ranchero* claim to a nonindigenous identity, making fun of the fact that, in spite of this claim, in this conversation they seem to be playing the Mexican government's game (we should value our indigenous heritage and perhaps know Tarascan). B has mentioned on other tapes how people at work question her claim that she is actually indigenous. Because she is so white, they tell her, she can't really be Indian. Her claim to those at work that she in fact is Indian can be interpreted as a response to the U.S. racial dichotomy of white/nonwhite. If she has any Indian blood, then according to this logic, she is not completely white, so she makes this claim herself rather than having this category imposed on her by others. She has told me that she is taken for white until she speaks (either Spanish, or English with a Spanish accent), and then has suffered job discrimination. Here, however, within the social network, she and D joke about these categories and invoke the shared knowledge that they are not truly indigenous according to the local logic of their micro-region within Mexico. There, the truly indigenous do speak *Purhépecha* and live in indigenous communities, excluding and being hostile to these "outsider" *rancheros*, who consider themselves to be of higher status. When, immediately after D's counterdiscursive *Pinche indios*, she playfully says, in 32, *Ehi, calmada con los indios, yo soy India* (hey, take it easy on the Indians, I am Indian), she ambiguously invokes both positions: first, in the United States, she might as well be Indian since she is nonwhite (conveyed by the literal meaning of her words), yet second, in Mexico, she knows she is not Indian, in spite of the government's official position (conveyed by the playful tone in which she expresses the words).

The joking on this topic begins to come to a close in 33 when D admits that they are all (ambiguously) Indian, like B, and then invokes the common metaphor for looking (very) Mexican: having a *nopal* cactus on one's forehead. They laugh and B ends the episode with an evaluative comment on those who would joke about such things in 36, *Ay, como son tremendas* (Oh, how audacious you all are). The word *tremenda* is often used positively, not pejoratively, by these women to refer to others who do not keep to "traditional" demure female behavior and are not afraid to speak up against such norms. B uses it here in that sense, showing her, and their, pleasure in this counterlanguage, so characteristic of *relajo*.

VERBAL PLAY AS RACIAL CRITIQUE

An extensive literature on joking reveals its capacity for social inversion (Bauman 1986, 70–77; Briggs 1988, 171–232), and I have argued (Farr 1998) that Mexican *relajo* functions as a micro-fiesta in this regard, since the fiesta, or carnival, is similarly an anti-structural process. In both verbal and nonverbal play such as *relajo* and fiesta, the usual norms and structures of society can be turned upside down, at least for the moment. Limón (1982), Bauman (1986), and Briggs (1988), among others, stress the creative and performative power of such play, especially verbal play, arguing for the transformative power of language, and that change is indeed facilitated by the critical perspectives engendered by joking (Farr 1994). In the above excerpt, B and D express a Bakhtinian "double-voicedness." One voice expresses the official Mexican ideology of pride in their Indian heritage, whereas the other voice is critical of this ideology, knowing full well the daily realities in the countryside that are linked to being Indian or *ranchero*. Their own movement back and forth across the Mexico–U.S. border and their familiarity with two different national ideologies that implicate race in different ways provides them with a perspective from which to critique these ideologies, a critique that is implicit in their joking. The ambiguity of their position in both national racial orders provides fertile resources for such humor. They consider themselves nonindigenous in Mexico, even as they are Indianized because they are rural peasants, and in the United States they are not easily placed in a single racial category. Although many of them are initially perceived to be white in the United States, they know they are not *güera güera* (really white) because of their mixed heritage. Neither are they black, nor Asian, nor Native American. Here, however, in this excerpt of joking, these women play with the ambiguity inherent in their not fitting neatly into either country's racial categories. Such critical verbal play, I would argue, can lead to an eventual realignment of traditional racial identities and attitudes, leading those who engage in this verbal play to question and undermine existing racial hierarchies.

CONCLUSION: RACE, *RANCHEROS* AND NATIONS

I have described the traditional racial ideology of *rancheros* in relation to colonial and postcolonial Mexican racial categories, as well as to U.S. racial categories and the place, or lack of it, of Mexicans within the traditional white/nonwhite dichotomy. Both Mexican and U.S. nation building have utilized the idea of shared descent and thus race in their imagined communities, Mexico with a new "mixed" race (*la raza cosmica*) and the United States with whiteness, ignoring differences in class, gender, and ethnicity in the attempt to essentialize national communities. According to Basch et al. (1994), transnational migrants have the potential to disrupt these homogenizing forces of nationalism. By resisting inclusion in either nation's racial categories, as well as by resisting the impositions of both governments, *rancheros* such as these affirm their difference from both nation's dominant identities. This in itself provides a space for counterhegemonic effects, as Basch et al. have pointed out:

> However, the issue of resistance is a complex one that must be contextualized within the always partial and unfinished construction of identities shaped by the pressures of national hegemonies. Subordinated populations may internalize many of the meanings and representations that pervade their daily surroundings, but that internalization remains partial and incomplete. Meanings are often subverted and there is always, at the level of daily practice, some opening for innovation. (Basch et al. 1994, 46)

These *rancheros*, and thousands of others, have "voted [in Mexico] with their feet" (Dinerman 1983) in migrating to the United States, where their increasing presence alone disturbs the traditional racial order. In their daily practices, which include *echando relajo* (joking around) as in the excerpt above, they resist both Mexican and U.S. hegemonic constructions of identity, playing with their ambiguous places in the racial orders of both countries. In daily verbal practices such as *relajo* they perform identities somewhere between the "us" and the "them" in both countries, illuminating the ground-level nuances of identities that don't easily fit discrete categories of a racial hierarchy.

NOTES

1. The term *indigena* ("indigenous") is used in Mexico to refer to those who are native to Mexico. "Indian" (*indio*), although it is sometimes used in the *rancho*, is considered less polite. *Güare*, or the diminutive *güarecita*, is a name used in the *rancho* to refer to indigenous women.

2. This house now has three bedrooms (the newer two bedrooms and bathroom having been built with money from Chicago). Like others of their generation, these parents, now in their fifties, had seven children now ranging in age from the late teens to the early thirties. The parents use one bedroom (a luxury not shared by all in the *rancho*), the female offspring use another, and the male offspring use a third. When more people are "home" from Chicago (including not only the two eldest daughters, but cousins and guests), females share the two double beds and extra mattresses put on the floor in the "female" bedroom. The father has remarked that it is all right for women to sleep all over each other, but not for men, so he plans to build more rooms onto the house.

3. Interview with Salvador Zambrano, February 5, 1996.

4. Interview with Aurora Carabes, June 30, 1998.

5. Until recently, public representations and most studies ignored another complexity in the racial history of Mexico: the presence of Africans. Aguirre Beltran (1946) pioneered the study of Africans in Mexico, but until recently, most have assumed that the presence of Africans was limited primarily to coastal areas. Highland Michoacan, however, including the micro-region of the present study, had significant numbers of Africans who were brought to Mexico as slaves to work in households, mines, and on sugar plantations (Chavez Carbajal 1994). This "third root" of Mexico has not been studied until recent decades because the historical awareness of African presence was buried as Africans assimilated, as individuals, into the population. Lomnitz-Adler (1992) explains that, whereas colonial policy allowed Indians a group identity within a hierarchical "Indian nation," Africans were not allowed to form groups that promoted a separate African identity. Slavery was justified as a transitory condition that enabled the Spaniards to convert individuals whose nations of origin rejected the faith. Thus individual Africans, but Indian nations, were "redeemed" as they converted to Catholicism.

6. In the 2000 Census, people were able to identify either as one race or as more than one. Although this is more reflective of contemporary reality, it still perpetuates the notion that a "race" is biologically real.

7. During a visit with relatives of this family in Mexico City, I was told that an educated (late) uncle had traced their roots to a town in Asturias, which they showed me on a map. The family name is the same as that of the town.

8. Elite urban Mexicans, for example, celebrate the Day of the Dead, an originally indigenous practice. *Rancheros*, however, generally do not celebrate this event, associating it with indigenous identity. Interestingly, some

younger members of the families in the present study began to celebrate this event in Chicago, where it has come to represent Mexican, rather than indigenous, identity.

9. This phrase is commonly used to mean that people appear to be native to Mexico, that is, they look Indian.

REFERENCES

Agnirre Beltran, G. 1946. *La población negra de México, 1519–1810*. Mexico: Ediciones Fuente Cultural, México.

American Anthropological Association Statement on Race. 1998. *American Anthropologist* 100, 3: 712–713.

American Association of Physical Anthropology Statement on Biological Aspects of Race. 1996. *American Journal of Physical Anthropology* 101: 569–570.

Ano Nuevo Kerr. 1977. Mexican Chicago: Chicano assimilation aborted, 1939–54. In *Ethnic Chicago*, M.G. Holli and P. d'A. Jones (eds.). Grand Rapids, MI: William B. Erdmans Publishing Co. 269–298.

Barragán López, E. 1990. *Más allá de los caminos*. Zamora: El Colegio de Michoacán.

Barragán López, E. 1997. *Con un pie en el estribo: Formación y deslizamientos de las sociedades rancheras en la construcción del méxico moderno*. Zamora, Michoacán, Mexico: El Colegio de Michoacán.

Barragán López, E., O. Hoffmann, T. Linck, and D. Skerritt (eds.). 1994. *Rancheros y sociedades rancheras*. Zamora: El Colegio de Michoacán.

Barth, F. 1969. *Ethnic Groups and Boundaries: The Social Organization of Culture Difference*. Boston: Little, Brown, and Co.

Basch, L., N. Glick Schiller, and C. Szanton Blanc. 1994. *Nations Unbound: Transnational Projects, Postcolonial Predicaments, and Deterritorialized Nation-States*. Amsterdam: Gordon and Breach.

Bauman, R. 1986. *Story, Performance, and Event: Contextual Studies of Oral Narrative*. Cambridge, UK: Cambridge University Press.

Bonfil Batalla, G. 1996. *México profundo: Reclaiming a Civilization*. Austin: University of Texas Press.

Briggs, C. 1988. *Competence in Performance: The Creativity of Tradition in Mexicano Verbal Art*. Philadelphia, PA: University of Pennsylvania Press.

Chávez Carbajal, Maria Guadalupe. 1994. Propietarios y esclavos negros en Valladohid de Michoacán, 1600–1650. Morelia: Universidad Michoacona de San Nioclás de Hidalgo, Instituto de Investigaciones Históricas.

Cochet, H. 1991. *Alambradas en la sierra*. Zamora, Michoacán, Mexico: El Colegio de Michoacán.

DeGenova, N. 1998. Race, space, and the re-invention of Latin America in Mexican Chicago. *Latin American Perspectives* 25, 5: 87–116.

Denton N. and D. Massey. 1989. Racial identity among Caribbean Hispanics: The effect of double minority status on residential segregation. *American Sociological Review* 54: 790–808.

Dinerman, I.R. 1983. *Migrants and Stay-at-Homes: A Comparative Study of Rural Migration from Michoacán, Mexico*. La Jolla, CA: Center for U.S.-Mexican Studies, University of California, San Diego.

Elias-Olivares, L. and M. Farr. 1991. *Final Report to U.S. Bureau of the Census: Sociolinguistic Analysis of Mexican-American Patterns of Non-Response to Census Questionnaires*.

Farr, M. 1994. *Echando relajo*: Verbal art and gender among *mexicanas* in Chicago. In *Cultural Performances: Proceedings of the Third Berkeley Women and Language Conference*. Berkeley: University of California. 168–186.

Farr, M. 1998. *El relajo como microfiesta*. In *Mexico en fiesta*, H. Pérez (ed.). Zamora, Michoacán, Mexico: El Colegio de Michoacán. 457–470.

Farr, M. 2000. *¡A mi no me manda nadie!* Individualism and identity in Mexican *Ranchero* speech. *Pragmatics* 10, 1: 61–85.

Farr, M. Forthcoming. *Rancheros in Chicagoacán*: Ways of speaking and identity in a transnational Mexican community.

Field, L. 1998. Post-Sandinista ethnic identities in Western Nicaragua. *American Anthropologist* 100, 2: 431–443.

Flores-González, N. 1999. Puerto Rican high achievers: An example of ethnic and academic identity compatibility. *Anthropology and Education Quarterly* 30, 3: 343–362.

Foster, G.M. 1967/1979/1988. *Tzintzuntzan: Mexican Peasants in a Changing World*. Prospect Heights, IL: Waveland Press.

Friedrich, P. 1977. *Agrarian Revolt in a Mexican Village*. Chicago: University of Chicago Press.

Friedrich, P. 1986. *The Princes of Naranja: An Essay in Anthrohistorical Method*. Austin, TX: University of Texas Press.

Frye, D. 1996. *Indians into Mexicans: History and Identity in a Mexican Town*. Austin, TX: University of Texas Press.

González and González, L. 1974. *San José de Gracia: Mexican Village in Transition*. Austin, TX: University of Texas Press.

González and González, L. 1995. The period of formation. In *A Compact History of Mexico*, D. Cosio Villegas (ed.). Mexico, DF: El Colegio de Mexico.

González Méndez, V. and H. Ortiz Ybarra. 1980. *Los Reyes, Tingüindín, Tancítaro, Tocumbo, y Peribán*. Municipal Monographs of the State of Michoacán. Morelia, Michoacán: Government of the State of Michoacán.

Holli, M. and P.d'A Jones (eds.). 1995. *Ethnic Chicago: A Multicultural Portrait*. Grand Rapids, MI: Wm. B. Eerdmans Publishing Company.

Knight, A. 1990. Racism, revolution, and *Indigenismo*: Mexico, 1910–1940. In *The Idea of Race in Latin America, 1870–1940*, R. Graham (ed.). Austin, TX: University of Texas Press.

Lazarre, J. 1996. *Beyond the Whiteness of Whiteness: Memoir of a White Mother of Black Sons*. Durham, NC: Duke University Press.

Limón, J. 1982. History, Chicano joking, and the varieties of higher education: Tradition and performance as critical symbolic action. *Journal of the Folklore Institute* 19: 141–166.

Lomnitz-Adler, C. 1992. *Exits from the Labyrinth: Culture and Ideology in the Mexican National Space*. Berkeley: University of California Press.

Massey, D. 1981. Hispanic residential segregation: A comparison of Mexicans, Cubans, and Puerto Ricans. *Sociology and Social Research* 65: 311–322.

Massey, D. and N. Denton. 1989. Residential segregation of Mexicans, Puerto Ricans, and Cubans in selected U.S. metropolitan areas. *Sociology and Social Research* 73, 2: 73–83.

Massey, D. and N. Denton. 1993. *American Apartheid: Segregation and the Making of the Underclass*. Cambridge, MA: Harvard University Press.

Nelson, C. and M. Tienda. 1997. The Structuring of Hispanic Ethnicity: Historical and Contemporary Perspectives. In *Challenging Fronteras: Structuring Latina and Latino Lives in the U.S.*, M. Romero, P. Hondagneu-Sotelo, and V. Ortiz (eds.). New York: Routledge.

Oboler, S. 1995. *Ethnic Labels, Latino Lives: Identity and the Politics of (Re)presentation in the United States*. Minneapolis, MN: University of Minnesota Press.

Omi, M. and H. Winant. 1994. *Racial Formation in the United States from the 1960s to the 1990s*. New York: Routledge.

Ortner, S. 1996. *Making Gender: The Politics and Erotics of Culture*. Boston: Beacon Press.

Padilla, F.M. 1985. *Latino Ethnic Consciousness: The Case of Mexican Americans and Puerto Ricans in Chicago*. Notre Dame, IN: University of Notre Dame Press.

de la Peña, G. 1984. Ideology and practice in southern Jalisco: Peasants, Rancheros, and urban entrepreneurs. In *Kinship Ideology and Practice in Latin America*, R.T. Smith (ed.). Chapel Hill, NC: University of North Carolina Press.

Peñalosa, F. 1995. Toward an operational definition of the Mexican American. In *Latinos in the United States: History, Law and Perspective, Volume I, Historical Themes and Identities: Mestizaje and Labels*, A. Sedillo López (ed.). New York: Garland.

Rodriguez, V.M. 1997. The racialization of Puerto Rican ethnicity in the United States. In *Ethnicity, Race and Nationality in the Caribbean*, J.M. Carrion (ed.). Rio Pedras: Institute of Caribbean Studies.

Rodriguez, C.E. and Cordero-Guzman, H. 1992. Placing race in context. *Ethnic and Racial Studies* 15, 4: 523–541.

Schriffrin, D. 1988. *Discourse Markers*. Cambridge, UK: Cambridge University Press.

Scott, J. 1990. *Domination and the Arts of Resistance: Hidden Transcripts*. New Haven: Yale University Press.

Skerry, P. 1993. *Mexican Americans: The Ambivalent Minority*. Cambridge, MA: Harvard University Press.

Taylor, P. 1933. *A Spanish-Mexican Peasant Community: Arandas in Jalisco, Mexico*. Berkeley, CA: University of California Press.

Telles, E.E. and E. Murguia. 1990. Phenotypic discrimination and income differences among Mexican Americans. *Social Science Quarterly* 71, 4: 682–696.

Warren, B. 1985. *The Conquest of Michoacán: The Spanish Domination of the Tarascan Kingdom in Western Mexico, 1521–1530*. Norman, OK: University of Oklahoma Press.

11

African American Vernacular English: Roots and Branches

John R. Rickford

I have crossed an ocean / I have lost a tongue / From the roots of the old / one / A new one has sprung.

Grace Nichols, *I is a Long Memoried Woman*

Introduction

I would like to set the mood and theme for this essay by presenting the first two verses of a beautiful but little-known song by Zulema Casseux, "*American Fruit with African Roots*" (the images in this song were enhanced by a number of slides presented at the symposium).

> We came from a distant land,
> Our lives already planned.
> We came in ships from across the sea,
> Never again, home we'd see.
> And now, we've become,
> American fruit, with African roots.
> Mmm, hmm, hmm, hmm, hmm, hmm.
> Our masters saw we worked from morn till night,
> Never given human rights,
> Though years passed, things remain the same,
> Children born with no last names.
> What is to become of these,
> American fruit, with African roots?
> Mmm, hmm, hmm, hmm, hmm, hmm.

Although Zulema Casseaux's song deals with African American *people*, we can extend it to the linguistic and cultural traits that they brought from Africa, and modify her question to read, "What has become of these American Fruit, with African Roots?" The prevalent scholarly view in the first half of this century—resurfacing in a more sophisticated form in recent years—is that African Americans have few if any special linguistic fruit, and that their African roots were destroyed by the devastating experience of slavery. The aptly named Krapp (1924, 192–193) argued, for instance, that "not a single detail of Negro pronunciation or Negro syntax can be proved to have other than an English origin." Crum (1940, 111), discussing the Gullah dialect of South Carolina and Georgia, posed the question, "Is it African or English?" and responded, "The answer is very positive: it s almost wholly English—peasant English of the seventeenth and eighteenth centuries, with perhaps a score of African words remaining." Linguists writing more recently have avoided putting their feet in their mouths quite this firmly. But the contention that African American Vernacular English is essentially British settler speech transplanted to America or Southern White Speech transported northward, continues to receive support (Davis 1971, 96; Poplack 2000), whereas Africanists and creolists are dismissed as "substratomaniacs" (Bickerton 1981, 48).

In this essay, fittingly presented in a symposium on "English and Ethnicity," I follow Herskovits (1941), Turner (1949), Stewart (1969), Dillard (1972), and Alleyne (1980) in arguing a contrary point of view: that African American language and culture has distinctive features (fruit)—which link it inexorably with the continent from which African Americans came, and with the synthesizing creolizing experience that they shared with their *brethren* and *sisteren* (as the Jamaican Rastas put it—see Pollard 2000) in the Caribbean.

Though most of my discussion is about African *roots* and Caribbean similarities, I'm fully aware that no living entity, least of all, a language, can remain in a new environment for four centuries without evolving. And I say a little toward the end of my essay about the vibrant *branches* that African American language and culture has developed in the United States, and about what it has taken from and contributed to other ethnic groups. My discussion, therefore, is about both continuities and innovations—roots and branches. Although my focus is on language, I'll begin with other cultural elements since no language exists in a vacuum, and these other elements attest richly to the distinctiveness of African American and especially Gullah ethnic identity.

THE SEA ISLANDS

African and creole ethnic roots are nowhere more evident than in the Sea Islands off the coast of South Carolina and Georgia. These islands—Johns Island, Edisto, Hilton Head, Daufuskie, and others that most Americans have never even heard of—have been described (by John Szwed of Yale University) as "the most direct repository of living African culture to be found anywhere in North America." The factors that make them so are partly demographic (Africans outnumbered Europeans by as much as ten to one in this former rice and cotton farming region), partly historical (slaves from Africa were brought there until 1858, compared with 1808 in most regions—see Wells 1967 on the historic 1858 arrival of the slave ship *Wanderer*, bearing 400 slaves) and partly geographical (the isolation of the islands—some still accessible only by boat). But the continuities with Africa and the Caribbean are numerous, and clear. When I first went out to the Sea Islands in 1969, as an undergraduate, the linguistic and cultural resemblances with my native Caribbean were so pervasive that I had a sense of returning *home*, and subsequent research has only deepened this impression.

Beginning with examples of folk culture, we might note, for instance, continuities between Africa and the Sea Islands in elaborate patterns of basket weaving (see Rosengarten 1994). These baskets, in their "weave pattern . . . coiling technique, manner of stitching, and . . . use within an agricultural framework" are similar to styles found in Senegal, Nigeria, Togo, Benin, and Ghana (Jones-Jackson 1987, 18). Examples can be found both in early-twentieth-century photos in Edith Dabbs's wonderful (1971) book, *Face of an Island*, and in Patricia Jones-Jackson's (1987) book *When Roots Die*, showing that the tradition is still alive and well on the Sea Islands. Jones-Jackson notes that children on the islands start to learn basket weaving around the age of six. Baskets from the famous Mount Pleasant and other coastal areas are sold daily in Charleston markets and along US Highway 17, just north of Charleston.

In a similar vein are distinctive patterns of woodcarving, such as the graveyard carvings and walking sticks from Georgia. As art historian Robert F. Thompson has noted, these indicate their African sources in their color, surface, pose, detail, facial posture, and subject. The Gullah carvings are comparable to the woodcarvings found among the Djukas—descendants of runaways from slavery—in Suriname, South America. Unlike basketry, however, woodcarving is no longer a widely practiced Sea Island art.

From the viewpoint of African and Caribbean continuities, the double-hitched skirts that one sees in early 1900s photos of Sea Island women in the field are also interesting. The second belt was reported to serve a dual purpose: (i) to lift the skirt off the ground to keep it from getting wet or dirty; (ii) to give the wearer extra strength, which was reported to derive from an "African superstition" (Dabb 1971). The wearing of double-hitched skirts was also common among Guyanese women doing strenuous work as shown in a late-nineteenth-century woodcut of Guyanese coal-carriers, homeward bound (reprinted in Rickford 1978, 195). Note in both cases, incidentally, the art of carrying large loads on the head without using hands—the baskets in each case supported by a headcloth or *kata*. This word in turn has good African (Twi *kata* = "covering") and Indian (Bhojpuri) etymologies, and is found in West Africa as well as in Antigua and Colombia (Moses 1978, 110). Salikoko Mufwene (personal communication) has told me that the word also occurs (as *kata* or *nkata*) in Kituba and Kiyansi and in some varieties of Kongo (again, for the cloth or grass ring placed between the head and the load).

Colombia and other Caribbean areas also resemble the Sea Islands in patterns of fishing, particularly in the kinds of nets used and the style of casting, which are similar from Colombia and the South Carolina Sea Islands to West Africa. Busnell (1973) reports a West African fishing pattern involving a symbiotic, cooperative relationship between man and dolphin, which is also found on the Sea Islands (compare Jackson et al. 1974). In this pattern, fisherman rap on the sides of their boats in open waters to attract porpoises, which circle and scare fish into tight circles. This makes it easier for the fishermen as they cast their nets, and easier also for porpoises, as escaping fish scurry back out to them after each cast.

In the category of material culture, again, is the use of the large mortar (a hollowed-out block of wood) and pestle (stick) for pounding cassava, plantain, corn, or rice, which is found both in West Africa and the Sea Islands—as shown in photos from Lydia Parrish's (1942) *Slave Songs of the Georgia Sea Islands*—and also throughout the Caribbean. And among the many funeral customs shared by the Sea Islands and the Caribbean—probably derived from West Africa—are the singing and story telling *wakes* and the tradition of passing babies over the coffin to discourage the deceased's spirit from coming back to haunt them.

The subject of haunts brings us naturally enough to the rich patterns of folklore that still exist on the islands, including the well-known "Brer Rabbit" stories (often compared to the Anasi stories of West Africa and the Caribbean) and the lesser-known stories of "hags" (evils spirits, often in the form of an old women) who enter homes and ride or suck people (especially babies) at night. Interestingly enough, one scholar (Nichols 1983) thought such hags and methods of catching them (scattering salt or sand for these compulsive creatures to pick up grain by grain or putting salt or pepper in their skins when they shed them) were characteristic of the Sea Islands and Liberia and Sierra Leone but not the Caribbean. This putative difference was thought to provide indirect evidence that the Sea Islands differed from the Caribbean in their demographic sources. But in Guyana, the *hag* as a folklore figure is alive and well—although it is known as an ole *higue*, not *hag*, because of a sound change from [a] to [ai] before [g] which we find in other words (e.g., *baig* for "bag").

Finally, there are various noteworthy patterns of music and religious worship, including spirituals, ring shouts (with rhythmic handclapping and circle dancing—see Rosenbaum and Rosenbaum 1998), prayers, and the call and response ("Amen!") tradition. These resemble patterns found in the Caribbean, and are syncretisms or mixtures of West African and European traditions (Herskovits 1941). As Rosenbaum and Rosenbaum note, [shaut] "shout" is an Afro-Asiatic word meaning "dance," and when it's done in the traditional way, the movement is counter-clockwise, as is the movement along the four points of the Kongo cosmogram, which I'll turn to next. Shouts are performed almost every Sunday afternoon in the context of usher marches on John's Island and other South Carolina Sea Islands.

One finds an intriguing configuration of holes—a diamond enclosing a cross—next to each of the supporting posts in the basement of the First African Baptist church in Savannah, Georgia, reputedly the oldest African American church in North America. The holes served as breathing holes for slaves who were secreted in the crawl space underneath the church on their way to freedom in the North via the "underground railroad." But their pattern, as a deacon from the church explained when we visited on an AAAS Learning Expedition in 1999, is that of a "Kongo Cosmogram." I had to go to the Web, to Robert Farris Thompson's *Flash of the Spirit*, and to Grey Gundaker's (1998) *Signs of Diaspora*, to find out more. As Gundaker notes, "The cosmogram is a visual summary of key concepts of Kongo cosmology [beliefs about the cosmos] that resonate with aspects of cosmologies among other peoples ranging from Angola northward to Edo OluKum worshippers in Nigeria." The cosmogram or dikenga may take several forms, but in its most common form it consists of a cross enclosed in a diamond or a circle. The extremities of the cross, sometimes ending in discs or circles, represent the daily [counterclockwise] movement of the sun, from its rising in the east, to its noontime peak, to its setting in the west and disappearance from view, to its darkest, southern point [midnight], to be broken again by the rising of the sun in the east. The four moments of the sun also mirror the Bakongo's belief in reincarnation and the continuity of likfe, from birth (E), to the peak of a person's life (N), to death (W), and passage through the underworld (S) until he/she comes back in the name or body of a child, or something else. The horizontal *kalunga* line separates the mountain of the living—*ntoto* (the triangular top of the diamond) from its mirrored counterpart, the inverted mountain of the dead or white clay, *mpemba*. God is imagined at the top, the dead at the bottom, and the watery kalunga line in between. As Thompson notes in his book, *Flash of the Spirit*, "This Kongo 'sign of the cross' [Yowa] has nothing to do with the crucifixion of the son of God, yet its meaning overlaps the Christian vision. The Bakongo did believe in a Supreme deity, Nzambi Mpungu, but they had their own notions of the indestructibility of the soul. The Kongo cross, [like many surviving Africanisms—Rickford and Rickford 1999] 'passes' itself off as a Christian symbol in a Christian setting, but it stands for the cyclical movement of human souls around its intersecting lines."

Two further comments about these cosmograms in the Georgia Sea Island area: (i) I showed Professor Thompson a copy of the slide I described to you just now, and he was fascinated. The world authority on this subject, he knew nothing of the perfectly preserved cosmograms in this Savannah church. There is a partial cosmogram on a broken brick on the Levi Jordan plantation in Texas (do a search for "Kongo Cosmogram" on Google or any other search engine, and it'll come up), but the ones in the First African church are numerous (they recur at every supporting post), and whole, perfectly preserved. (ii) Kongo cosmograms are found elsewhere in the African Diaspora, including Jamaica, at the Sam Sharpe monument in Heroes Park, Kingston, which I visited on another African and African American studies Learning Expedition in 2000. They are also found in Cuba and Brazil, where there were heavy importations of Kongo and Angolan slaves. As in the Kongo, they are often found on the ground, marking a sacred spot on which one stands to make an oath, or sing, and draw a sacred point.

Striking as these varied patterns of folklore, folk culture, and religious belief and practice are, the richest fruit on the Sea Islands— from the viewpoint of a linguist, at least—is its distinctive Gullah dialect, the creole or vernacular English of the region. As I noted earlier, late-nineteenth and early-twentieth-century scholars had felt that Gullah had few if any African roots—maybe a score of African words at most. But Lorenzo Dow Turner—one of the earliest African American linguists—spent *fifteen* years doing fieldwork and library research that *blew* these earlier misconceptions away. In his 1949 book, *Africanisms in Gullah Dialect*, Turner listed 4,000 Gullah words that had plausible African etymologies. Most of them were personal nicknames (also known at *pet names* or *basket names*) used among family and friends, in contrast to English names, which they used at school and with strangers (like earlier Gullah scholars, who did not remain on the island long enough to get beyond formalities). Two examples from Turner's extended list are *aba* and *abako*. In the case of *aba*, one source is the Fante (Gold Coast) word a_3ba_3, a personal name corresponding to the Twi *ya*, "name given to a girl born on Thursday."

In addition to the personal nicknames (discussed in Mufwene 1985) Turner also found about 400 other African derived words used in daily conversation. Some of these such as *tote* and *gumbo*, are widespread throughout the United States, but you may not have been aware of their African pedigree. Others, such as *bakra* and *nana* ("white man" and "grandmother" respectively) are better known in the South and the Caribbean.

Turner also found a smaller but intriguing set of African forms used in prayers, stories, and songs, such as the words na_3na_1, $tu{:}_1$, and $gbang_3$ (with West African coarticulated *gb* stop) in this Vai-Gullah song: "New rice and okra / na_3na_1, na_3na_1 / Beat rice tu${:}_1$ gbang$_3$, gbang$_3$ / na_3na_1, na_3na_1."

This last song, mixing as it does both English and West African words, will serve as a good entrée to the further point that Turner wanted to make: that Gullah was grammatically similar to both African languages and to Afro-English dialects spoken in West Africa and the Caribbean. When speakers of different languages develop a mixed, often simplified lingua franca to communicate among themselves, it is known as a pidgin. When that pidgin is learned natively, for instance, by children of Yoruba and Ibo slaves born on a South Carolina plantation, it expands in grammatical resources to meet the need of a primary language, and is known at that stage as a creole. One of the striking features of such creoles, particularly in the Atlantic, is how *much* they resemble each other. Although Louisiana creole and Gullah use words from different European languages (French and English respectively), they are similar partly because of common West African *grammatical* roots.

To exemplify some of the grammatical structures of Gullah, I introduce some Gullah speakers and discuss significant sentences that they used.

Let's start with Wallace Quarterman, born a slave in Georgia in 1844, who was an invaluable source of information to both Parrish and Turner in the 1930s. The following sentence is from a 1935 interview recorded by Alan Lomax, Zora Neale Hurston, and Mary Elizabeth Barnicle that was gathering dust in the Library of Congress until it was rediscovered recently (See Bailey et al. 1991). Quarterman has just described the moment when the "big gun shot"—when a fleet of fifty Union vessels steamed into Port Royal, firing on weak Confederate forces on Hilton Head and Bay Point and signaling the end of slavery—the "day of Jubliee"—for the slaves. In his gripping first-person account of this historic moment, Quarterman states that he relayed a message from the massa to the overseer in the fields to "turn the people loose." After this, in his dramatic words:

> de people dem t'row 'way dey hoe dem. Dey t'row 'way dey hoe, an' den dey call we all up you know an', an' gi' we all freedom 'cause we jus' as much free as dem.

Two linguistic features of note in this extract are: (i) the use of a third person plural pronoun form (*dem*), suffixed to the noun to form the plural (*dey hoe dem*); and (ii) the non-differentiation of pronouns used as subject, object, and possessive: *dey* hoe, gi' *we*, instead of the case marked forms "their" and "us" respectively. Both features are found in Jamaican, Guyanese, and other Caribbean creoles, and resemble grammatical features found in Yoruba, Ibo, and Ewe (Turner 1949, 223–229).

The next speaker I'd like to discuss briefly is Mrs. Queen, whom I interviewed myself in 1970. She was born around 1886. One of the many linguistic riches I got from her was this sentence:

> My aunt useta live in Washington, wa build da house over dey, da house wa Rufus *de* in.

The italicized *de*—equivalent to "is," but used only in locative phrases—seems at first to be nothing more than a reduced from of the English "there," but it has more plausible West African etymologies in Twi and Ewe, where there's a verb *de* with the meaning or function of "be" (Allsopp 1996, 188; Cassidy LePage 1980,144–145).

The next speakers are a couple, whom I refer to as Mr. and Mrs. Hope (also interviewed by me in 1970, and like Mrs. Queen, now deceased). They were financially poor, but linguistically rich. Here's a sample of

Mrs. Hope's speech:

> Yeah, he *does be* up an' Ø cut wood sometime, an'Ø go in de wood, Ø
> get lil wood and all. And he Ø use dese muss (moss) fuh
> tobacco. . . . He *does use* dese moss. [II:266–268]

Note in particular her use of an unstressed *does* ("He *does* use dese muss") to encode habitual aspect, the fact that something happens regularly. Turner's 1949 book did not include any examples of this form, only of a *da* form, as in:

> dem ca' um gi'de young people wuh da wuk dey (265)
> (They carried it for the young people who usually work here)

Now I encountered a few examples of this *da* (which Turner, 213–214 linked to African sources) in the 1970s, but *does* was more frequent. *Does* was itself on its way out—it was used only by the oldest speakers, and even they typically used it only in reduced forms, like *oes*, and *s*, and sometimes they left it out, as in the sample above. The Sea Island kids did not use *does* at all, so that where Mrs. Hope might alternate between "He does use dis moss" and "He Ø use this moss," the children used only "He Ø use this moss." I interrupt this saga of changes for now, but I return to it presently, because it forms a backdrop without which the development of forms in urban African Vernacular English cannot be fully understood.

U.S. Mainland-Inner Cities

When it comes to the language and culture of African Americans on the U.S. Mainland, particularly in the urban areas of the north and west, the belief that African and/or creole roots have been lost is even more pervasive. Clearly, Oakland, California, or St. Paul, Minnesota, are not the same as Daufuskie Island, South Carolina, or Harris Neck, Georgia. But the African American populations in these cities are not as rootless and ahistorical as some would have us believe; there *are* connections between the American cities and islands, and between them both and the Caribbean and Africa. They are not usually evident on the surface, however, and require an informed comparative perspective to reveal themselves.

Let's take, for instance, the case of invariant habitual *be*, as in this sample of speech from Foxy Boston, a thirteen-year old girl from East

Palo Alto, California, whom we recorded a few years ago. Here she is talking about her dreams:

> I *be* wakin' up an I be slurpin, I *be* goin'. DANG, THA'S SERIOUS!

The uninflected *be* in these sentences is one of the most distinctive features of urban African American Vernacular English—and is adequately translatable into Standard English only with some awkwardness. For instance, the slogan that Arsenio Hall, the black talk show host, used to use on his show, "Arsenio Hall—we *be* having a ball!" would be, in Standard English, "Arsenio Hall—we are usually having a ball!" (That just doesn't ring right, does it?). Well for years, people had wondered where this uninflected habitual *be* came from, but no convincing answers had emerged. Irish and Irish-English had been mentioned as possible sources, but the process of transmission and development remained problematic.

However, one of the ongoing changes on the Sea Islands that I was able to witness in the early 1970s was the emergence of *be* as Gullah's primary habitual marker. This seemed to me to be a recapitulation of a process that had happened earlier and more generally on the U.S. Mainland, and I was able to hypothesize a series of interrelated steps in the process, which helped to solve the puzzle of *be*'s origin.

You may recall from what I said about the Sea Island verb forms earlier that their former habitual marker, *da*, had changed to *does*, and that *does* itself was being reduced to zero. Well, while all this was going on, the *de* that the very oldest speakers also used in certain sentences, for instance between *does* and a preposition, had also changed more generally to *be*. When *does* was finally eroded, this *be* form came to mark the habitual meaning formerly carried by *does*. It may sound a bit confusing, but the following example will help to summarize and clarify the process:

a. He *da de* up an cut wood sometimes

b. He *does de* up and cut wood sometimes (da → does)

c. He *does be* up and cut wood sometimes (locative de → be)

d. He Ø *be* up and cut wood sometimes (does → Ø)

Stage a is attested only in books from earlier periods, like Turner's; the older speakers whom I recorded in 1970 showed evidence of stages b and c and occasionally d; the youngest speakers were almost exclusively d. Overall, the development of *be* is a part of the general process of decreolization—evolution *away* from creole roots *toward* the forms and conventions of Standard English—which has been going on in other areas of Gullah grammar, and in the Caribbean creoles. Several other central features of African American Vernacular English represent decreolization at work, as we'll see later. One case in which the African and creole roots of inner-city forms are visible only through their Sea Island and Caribbean connections involves the use of *say* as a complementizer, equivalent to Standard English "that," as in this sentence which I recorded from a thirty-one-year old man in Philadelphia:

> They told me *say* they couldn't get it in time.

Now at first it's easy to interpret this as derived from English *say* (equivalent to "They told me, they said"). But on the Sea Islands, as in the Caribbean, the complementizer is always *say*, uninflected, and it's used, not only with *tell*, but also with verbs like *know*, *think*, and *believe*, which refer to cognitive processes, as in Sarah Grant's comment:

> You wouldn believe *say* is a colored woman own that house.

As it turns out, Twi and other West African languages use a native form *se* as a complementizer (Allsopp 1996, 489; Cassidy 1961, 63; Dillard 1972, 121), much as the Gullah and creole speakers do, as in this Twi sentence (Turner 1949, 211):

> @nna o susuwi *se* eye okramang foforo bi.
>
> Then he thought that [se] it was some other dog.

The case of *say* then, must be treated as a case of multiple etymology or convergence, where African sources are masked behind formally similar English forms (Cassidy 1961).

As Angela Rickford and I emphasized in a 1976 article (Rickford and Rickford 1976), *masked Africanisms* such as *say*—which can "pass" as English forms—are important because they are more likely to have survived centuries of acculturation to European languages and Eurocentric prejudices than direct African retentions such as *nyam* "eat"—which is now found only in the Sea Islands and the Caribbean, not in U.S. cities.

Having given you such a heavy dose of linguistics, I want to turn away from language now and return to folklore as my final example of

West African and Caribbean connections with the U.S. mainland. Have any of you read about slaves who could fly? Slaves flying back to Africa?

Well, the motif of Africans and African Americans who could fly has received little attention in the *scholarly* literature to date. I wrote a little about it in my 1987 book *Dimensions of a Creole Continuum*. Some of you may have come across this theme, however, in fiction, as in Toni Morrison's novel *Song of Solomon*, or in poetry (Robert Hayden), or in folklore, as in Virginia Hamilton's beautiful book *The People Could Fly*. All of the examples so far are from the U.S. *mainland*.

But I first encountered this story myself on the South Carolina Sea Islands, from a middle-aged schoolteacher who told me that *her* great-grandparents had flown back to Africa after receiving a whipping. She looked me dead in the eye as she told me this, to make sure I didn't smile. And I didn't.

Subsequently, I came across dozens of other references to this motif—in *Drums & Shadows*, a 1940 book based on interviews with former slaves and other old people from South Carolina and Georgia. And in the fascinating *Autobiography of a Runaway Slave*—Miguel[Migel] Barnet's book about Esteban Montejo, a Cuban slave who didn't realize slavery was ended until he was found, hiding in the forests, some time later. Esteban said that African slaves didn't escape slavery by committing suicide, as some said, but by *flying*: "They flew through the sky and returned to their own lands." He added further details, even: that the Musundi Congolese flew the most, and that they disappeared by means of witchcraft—by fastening a chain to their wrists, which was full of magic. "I knew all this intimately," he said, "and it is true without a doubt."

The connection with magic and ritual was reinforced in a story told to me in my native Guyana a few years later by Damon, then in his seventies. He said that *his* great-grandfather, a former slave from the Popo tribe, told him that when they were coming from Africa, slaves who were brought to the top decks for fresh air would form a circle and begin to chant and sing African songs. And as the rhythm grew stronger and stronger, they just took off and *flew*! After a few occurrences of this, the ship crews reportedly kept them below decks to ensure that they arrived at their New World destinations.

The motif of slaves who could fly shows up also in Jamaica, and in Suriname (formerly Dutch Guiana), as depicted in a portrayal of slaves with wings, from Petronella Breinburg's book of Suriname legends (referred to me by the late, great St. Clair Drake). Interestingly enough, a different connection with ritual is established there—slaves who had eaten salt pork (typical of a slave diet) reportedly lost the magical ability to fly and plunged to their deaths.

So now we have established that the depictions of people flying in Morrison's novel or Hayden's poem or the schoolteacher's family lore are not just the products of their individual imaginations, but evidence of a theme that runs throughout the African New World—*African America*, if you will, in the broad sense that includes Cuba and Jamaica, and Guyana, and Suriname.

What is the deeper significance of this motif? Well it seems to have connections with *spirit possession* (when people, as a result of similar incantations and ritual preparations, are able to symbolically or spiritually leave their usual selves and take on the spirits of others, living and dead). And the concept of human flight is paralleled in other aspects of African American folklore, as in the case of the *hags* or *higues* who fly through the air in search of victims. Although we have so far found no exact analogs in West African folklore (we have barely begun to search), the African belief in the transmigration of the soul to the place of one's birth after death may be at the base of this recurrent theme of "flying back to Africa." More likely, however, this is one of those originally African elements that has been remodeled or reinterpreted by the experience of the Middle Passage and slavery, maybe providing a symbolic release from experiences that one could transcend in no other way. Certainly slaves who jumped into the sea during the Middle Passage, or who were beaten to death, would, by the doctrine of the transmigration of the Souls, return to Africa. But the existence of this possibility of flight—like the related belief that slaves who were being beaten could have the hurt telepathically transferred to the slave master's *wife* (Warner-Lewis 1991, 1996)—was one of the few aces that slaves had up their sleeves. It's significant in the stories that they often pulled these aces when the harshness of their New World experiences became most pronounced—for instance, after a whipping—from which it provided symbolic relief.

BRANCHES / INNOVATIONS

Now let me address the topic of innovations as well as continuities, about branches or shoots as well as roots, concentrating once again on language, my specialization.

While *be* has African and creole roots in *does be, da de*, and similar forms, it is unique in developing into the primarily habitual marker as it has, and also in developing still further now, into primarily an auxiliary, before progressive verbs as in *I be walkin* versus *I be sick* (adj.) or *I be at home* (location). That's a complex and ongoing process that I can't discuss in detail now, but it has engaged the attention of other linguists (e.g., Bailey and Maynor 1987).

The other innovations in AAVE that are most striking are also in the preverbal or auxiliary slot:

1. Stressed BIN as marker of remote aspect or phrase "I BIN had that" = "I've had that for a long time"—NOT in the creoles.
2. Combinations of *be* and *done* to express the Future Perfect and Conditional: "Teena better watch out—she *be done* took over."
3. *Come* as a modal auxiliary, to mark the speaker's indignation: "He *come* tellin me to move as if he own the place."
4. *Finna* (from *fixin'* to) as a marker of immediate future, as in "We finna go"—present to be sure, in Southern English, but being used more extensively by northwestern African Americans, especially in this reduced form.

None of these forms is used in quite the same way in Caribbean or West African pidgins and creoles, although Bin and *come* at least have some parallels. They testify to the fact that language is constantly changing—African American Vernacular English no less than any other.

INFLUENCE FROM/TO OTHER GROUPS

Although I have stressed *African* continuities in this essay, African American language and culture has clearly borrowed from and given to other groups.

One group whose links to African Americans remain to be fully explored is the Irish. The two ethnic groups were not close in the nineteenth century—when, although both groups were linked in "For Rent" signs that said: "No Blacks Nor Irish Need Apply"—they were competing for jobs and other scarce resources in an open economic market, and often rioted against and attacked each other. But they *were* close in the seventeenth century, when they were both bond servants in the closed plantation environments of the Caribbean and colonial America—almost equally despised by the Protestant English. In those days, they would rebel together, escape and hide out in caves together, and it seems likely that forms such as habitual *does* and *does be* could have entered the creoles from Irish English itself (see Rickford 1986).

In the lending category, examples are very numerous. Labov and others have shown, for instance, that Puerto Ricans in New York City have assimilated a lot of the vernacular language of African Americans, and runaway African slaves who went to live with the Seminole Indians in Florida transmitted to them their Gullah dialect—which their Afro-Seminole descendants in Texas still speak.

Examples from other aspects of popular culture—such as the hand-slapping and high fives in sports, or the influence of African Americans on jazz, abound. One example from a different genre of music—spirituals—that I discovered recently is the Sea Island "Shout"—"I Gotta Move" as sung originally by the Georgia Sea Island Singers (Carawan and Carawan 1994). The same song, only slightly transformed, was sung and recorded subsequently by Mick Jagger and the Rolling Stones (with no credit to the Sea Island source, of course).

CONCLUSION

Using thirty or more specific examples from the South Carolina Sea Islands and the U.S. mainland, I have tried to demonstrate that several aspects of African American language and culture are rooted in African traditions and paralleled in the creole language and culture of the Caribbean. Some of the continuities—such as the 4,000 African words in Gullah, or the almost exact equivalence of their baskets and mortars and pestles—are obvious, although outsiders to the Sea Islands barely know of their existence.

More subtle, more challenging, and intellectually more interesting are the continuities that exist on the mainland precisely because they have masked themselves in English [or Judeo-Christian] guise—such as *suck teeth* and the *say* of "He tell me say he gone" [or the Kongo cosmograms of the First Baptist Church in Savannah, Georgia]. In this category, too, are the seemingly trivial forms such as habitual *be*, or stories of slaves flying back to Africa, which we are tempted to treat as family folklore or the products of individual imagination. Their development from African and creole roots only becomes evident when we look at them closely, from a comparative perspective, and realize that these vessels of sound and meaning convey more than referential meaning—they are silent carriers of *history*.

I have also tried to draw your attention to some innovative aspects of African American language and culture (such as stressed *BIN* and *finna*), and of some respects in which it has borrowed from and lent to other ethnic groups (e.g., in music). These are the vibrant *branches* of African American language and culture.

In closing, let me return to the words of that song by Zulema Casseaux with which I began this essay: American fruit, with African roots. Some of you may have been wondering why I might draw on this metaphor for my essay. Why "fruit"? For there is a tendency in some quarters to regard some or all of the examples of African American vernacular language and culture I have described with shame, and to deny or deprecate them, especially in public. But they are systematic, they are rooted in history, and they serve to express the ideas, social relationships, and ethnic identities of thousands of African Americans daily. In these various respects they are certainly fruit— worth cultivating, worth sharing, and worth feeding on—for sustenance and health. As Grace Nichols says in the epilogue to her wonderful opbook of poetry, *I is a long memoried woman*, "I have crossed an ocean; I have lost my tongue; from the roots of the old one, a new one has sprung!"

We came from a distant land,
Our lives already planned.
We came in ships from across the sea,
Never again, home we'd see.
And now, we've become,
American fruit, with African roots.
Mmm, hmm, hmm, hmm, hmm, hmm.

Our masters saw we worked from morn till night,
Never given human rights
Though years passed, things remain the same,
Children born with no last names.
What is to become of these
American fruit, with African roots?
Mmm, hmm, hmm, hmm, hmm, hmm.

REFERENCES

Alleyne, Mervyn. 1980. *Comparative Afro American*. Ann Arbor: Karoma.

Allsopp, Richard. 1996. *Dictiionary of Caribbean English Usage*. Oxford: Oxford University Press.

Bailey, Guy and Natalie Maynor. 1987. Decreolization? *Language in Society* 16: 449–473.

Bailey, Guy, Natalie Maynor, and Patricia Cukor-Avila, 1991. *The Emergence of Black English: Texts and Commentary*. Amsterdam and Philadelphia: J. Benjamins.

Bickerton, Derek. 1981. *Roots of Language*. Ann Arbor: Karoma.

Busnell, R.G. 1973. Symbiotic relationship between man and dolphins. *Transactions of the New York Academy of Sciences* Series II, 35: 112–131.

Carawan, Guy, and Candie Carawan 1994. *Ain't You Got a Right to the Tree of Life? The People of Johns Island, South Carolina, Their Faces, Their Words, and Their Songs*. Athens, GA: University of Georgia Press.

Cassidy, F.G. 1961. *Jamaica Talk*. New York: St. Martin's Press Inc.

Cassidy, F.G. and R.B. Lepage. 1980. *Dictionary of Jamaican English*. New York: Cambridge.

Crum, Mason. 1940. *Gullah: Negro Life in Carolina Sea Islands*. Duke University Press.

Dabbs, Edith M. 1971. *Face of an Island: Lee Richmond Miner's Photographs of St. Helena Island*. New Year: Grossman.

Davis, Lawrence M. 1971. Dialect research: Mythology and reality. In *Black–White Speech Relationships*, Walt Wolfram and Nona Clarke (eds.). Washington, DC: Center for Applied Linguistics. 90–98.

Dillard, J.L. 1972. *Black English: Its History and Usage in the United States*. New York: Vintage Books.

Gundaker, Grey. 1998. *Signs of Diaspora: Diaspora of Signs*. New York: Oxford University Press.

Herskovits, Melville J. 1941. *The Myth of the Negro Past*. Boston: Beacon Press.

Jackson, Juanita, Sabra Slaughter, and J. Herman Blake. 1974. The Sea Islands as a cultural resource. *The Black Scholar* 5, 6: 32–39.

Jones-Jackson, Patricia. 1987. *When Roots Die: Endangered Traditions on the Sea Islands*. University of Georgia Press.

Krapp, George Philip. 1924. The English of the Negro. *American Mercury* June: 192–193.

Montejo, Esteban. 1968. *The Autobiography Of A Runaway Slave*. New York: Pantheon Books.

Moses, Lloyd. 1978. Cane-farming terms in Guyana. In *A Festival of Guyanese Words*, 2nd ed., J. Rickford (ed.). Georgetown: University of Guyana. 101–110.

Mufwene, Salikoko. 1985. The Linguistic Significance of Africa proper names in Gullah. *New West-Indian Guide* 59, 1: 46–66.

Nichols, Grace. 1983. *I Is a Long Memoried Woman*. London: Karnak.

Parrish, Lydia. 1942. *Slave Songs of the Georgia Sea Islands*. New York: Creative Age Press.

Pollard, Velma. 2000. *Dread Talk: The Language of Rastafari*. Montreal, Canada: McGill-Queen's University Press; Kingston, Jamaica: Canoe Press.

Poplack, Shana (ed.). 2000. *The English History of African American English*. Oxford: Blackwell.

Rickford, John R. (ed.). 1978. *A Festival of Guyanese Words*, 2nd ed. Georgetown: University of Guyana.

Rickford, John R. 1999 [1986]. Social contact and linguistic diffusion: Hiberno English and New World Black English. In *African American Vernacular English*, J.R. Rickford (ed.). Oxford: Blackwell. 174–218.

Rickford, John R. and Angela E. Rickford. 1999 [1976]. Cut-eye and suck–teeth: African words and gestures in New World guise. In *African American Vernacular English*, J.R. Rickford (ed.). Oxford: Blackwell. 155–174.

Rosenbaum, Art and Margo Newmark Rosenbaum. 1998. *Shout Because You're Free: The African American Ring Shout Tradition in Coastal Georgia.* Athens and London: The University of Georgia Press.

Rosengarten, Dale. 1994 [1986]. *Row upon Row: Sea Grass Baskets of the South Carolina Lowcountry.* McKissick Museum, University of South Carolina.

Savannah Unit, Georgia Writers Project, Work Projects Administration. 1986 [1940]. *Drums and Shadows: Survival Studies among the Georgia Coastal Negroes.* Athens, Georgia: Brown Trasher Books, University of Georgia Press.

Stewart, William A. 1969. Historical and structural bases for the recognition of Negro dialect. In *20th Annual Round Table*, James E. Alatis (ed.), Monograph Series on Language and Linguistics, no. 22. Washington, DC:Georgetown University Press., 239–247.

Thompson, Robert Farris. *Flash of the Spirit.*

Turner, Lorenzo Dow. 1949. *Africanisms in the Gullah Dialect.* Chicago: Unversity of Chicago Press.

Warner-Lewis, Maureen. 1991. *Guinea's Other Suns: The African Dynamic in Trinidad Culture.* Majority Press.

Warner-Lewis, Maureen. 1996. *Trinidad Yoruba: From Mother Tongue to Memory.* University of Alabama Press.

Wells, Tom Henderson. 1967. *The Slave Ship Wanderer.* Athens, GA: University of Georgia Press.

12

RACE AND ETHNICITY IN THE ENGLISH-SPEAKING WORLD

Janina Brutt-Griffler

The themes that emerge from the chapters of this volume fit well within the scope of the "Signs of Race" series, which "examines the complex relationships between race, ethnicity, and culture in the English-speaking world." Focusing on the centrality of the English language itself—and its increasing contact with other languages—the collection of essays employs approaches drawn from the field of linguistics. Its framing, *English and Ethnicity*, is in keeping with what Davies notes in her introduction to be linguists' preference for exploring issues of *ethnicity* over those of *race*, though the latter is a subject of growing interest in the field (cf. Makoni et al., 2003). In so doing, it offers interesting insights into another series aim, "exploring the ways in which race remains stubbornly local, personal, and present," and an implicit critique of the meaning and use of *ethnicity* as an analytical tool in the field of linguistics. In this conclusion, I would like to draw together the strands that emerge from these contributions and suggest how they provide directions for future work in the field.

The essays that comprise this volume, in investigating the relation of English and ethnicity, reveal complex, dynamic, shifting, and conflicting social and personal imperatives that belie the simplifying and truncating binary terms in which much of the contemporary politics of English has often been presented (for a review, see Brutt-Griffler 2002). Consider, for example, Mazrui's discussion of "English [as] a language of pan-African communication," p. 63 and its corollary, "English as exit visa" p. 68 from ethnic identity.[1] In Africa, he reports, "women [have been] more highly motivated to learn the language because it accorded them new opportunities to escape from their ethnically

ascribed status on grounds of their gender" p. 66. In this attempt
to "relocate themselves culturally," African women are "challenging the ethnically defined patriarchal boundaries of their identities
in new ways" p. 67. Mazrui also describes how gays and lesbians in
Africa have similarly "found English a useful facilitative tool in
their quest to live a gay identity," p. 67 resulting in "a critical role
for the English language in the interplay between ethnicity and
homosexuality in urban Anglophone Africa" p. 68. The value of
English, however, need not be limited to providing "an avenue of
escape from certain cultural constraints of one's ethnic group"
p. 68. It may also result in the transformation of that ethnic identity. The African Diaspora of imperialism, as Mazrui calls it, to the
United States results in the creation of American Africans, who in
turn undergo a transformation into African Americans "at precisely
the point when they lose their ancestral languages and acquire the
English language instead" p. 69.

Canagarajah's work highlights a parallel process in the Tamil Diaspora.
He notes within it a marked tendency toward a language shift to "English
monolingualism," prompting fears that the "Tamil language [is] going to
'die' within the community in the West and that the Tamil identity may
get erased within the next fifty years or so" p. 191. Indeed, his ethnographic research finds strong support for the first process: children born
of Tamil parents but raised in the West "are overwhelmingly monolingual
in English" p. 192. Gender once again emerges as a factor, as young
females in the diaspora attach even greater importance to English, up to
and including English monolingualism, than their male counterparts.
Canagarajah reasons, "It is possible that Tamil females are enjoying a new
sense of freedom and individuality that women haven't experienced traditionally in the Tamil community. Perhaps women are taking to English
more enthusiastically as it provides alternate identities that favor their
interests in the new life in the West" p. 195. He has also uncovered a similar phenomenon related to class, with "the less privileged families [using]
English to construct new identities in the new land" p. 194. As to the
maintenance of Tamil ethnic identity, a more nuanced picture emerges.
His subjects express the belief that they will maintain "a sense of
Tamilness that constitute[s] their identity" p. 197. Still, "Asked in general
if they would identify themselves as Tamils, the youth said that this identity was irrelevant now. They declared that they would identify themselves
as British or Canadian or American" p. 197. Canagarajah admits, "They
surprised me by questioning the exclusivist assumptions of my survey
questions. They argued that identity should not be based on all-or-nothing constructs—i.e., American or Tamil" p. 197.

One of Canagarajah's informants remarks, "We should keep back our own languages and speak one common language if we are going to join the mainstream life here" p. 195. In marked contrast, Toribio finds a completely opposite impulse among some Dominicans in the United States who "deploy a stigmatized variety (vis-à-vis Peninsular Spanish) of a stigmatized language (vis-à-vis English) in binding themselves to their Dominican compatriots and isolating themselves from their African and African-American neighbors" p. 133. She concludes, "U.S. Dominicans can simultaneously confirm and contest the identities foregrounded in the wider socio-cultural frame as well as project new identities in (re)constructing socio-cultural contexts" p. 148–149.

It is a recurrent theme in this volume that ethnicity is undercut, bounded, or even eclipsed by race, class, gender, and sexual orientation. Baugh, for example, divides children in the American school system not along ethnic lines, but into those for whom Standard English is native and those for whom it is not—a dichotomy in which usual notions of American ethnicity (white versus "minority") are significantly disrupted by class ("poor" versus "affluent": in "affluent homes . . . Standard English is the norm") p. 224. Coupland's study similarly revolves around the category of class rather than ethnic affiliation. Race constitutes at least an implicit theme in not only the contributions of Rickford, Baugh, and Mazrui, who to a greater or lesser extent concern themselves with African American Vernacular English, but those of Toribio and Farr as well, whose ostensible concern is with Hispanic Americans. Farr demonstrates that the Mexican (American) context can be as racialized as the Dominican, as evidenced by the racialized discourse of *güeros*, *prietos*, and *morenos*, none of which designations have ethnic significance. She even argues for a construction of *mestizaje* in explicitly racial terms.

One of the few to treat an ethnic variety of English, Huang's essay is paradigmatically innovative in surprising ways. For whereas, as Coupland notes, sociolinguistics is sometimes equated with a " 'variationist' or 'Labovian' or 'socio-phonetic' or 'secular' tradition," in Huang's view, what he terms Chinglish is an "invented vernacular" that "exists only as a literary language" p. 96. Moreover, it is one viewed by many Asian Americans as tainted by its association with "capitulation to the stereotype imposed on Asian Americans, as weak-minded, incompetent speakers of English" p. 96. Together with Baugh's discussion of African American Vernacular English, about which he notes, "African Americans have strong and diverse opinions," p. 221 it reminds us that the politics of variety usage (in this case a "translocal dialect") are, like the politics of identity, more complicated than often portrayed.

There is also at least an undercurrent of the influence of nonnative English speakers on the language. In fact, Huang's interest in Chinglish is motivated by its "critique of English . . . from within," p. 97 Chinglish as a language that defies what is often called native speaker ownership (Widdowson 1994) by being "different and exploratory" p. 97. Huang refers to his own personal "tinker[ing] with Shakespeare's language," p. 98 and quotes Henry James's reference to in-migrants to the United States who "play, to their heart's content, with the English language, or in other words, dump their mountain of promiscuous material into the foundations of the American" p. 99. The same theme finds echoes in Ortiz's essay, which not only describes code mixing, but also employs it.

To be sure, we find within this volume contributions representative of more traditional themes, including ethnic language varieties (Rickford, Huang, Coupland, Bernstein). Rickford's contribution, in particular, makes the case for the meaningful connections that exist between them, the speech communities that use them, and the historical processes that constitute and preserve them and provide the grounding for their further development. Rickford writes "about both continuities and innovations—roots and branches" p. 260. Within his "focus . . . on language," p. 260 he gives considerable attention to "other cultural elements since no language exists in a vacuum, and these other elements attest richly to the distinctiveness of African American and especially Gullah ethnic identity" p. 260. In a sense, African American Vernacular English is the classical case that sociolinguistics has traditionally had in mind. For example, in his widely used text, Trudgill (1995) opens the chapter entitled "Language and Ethnic Group" with a discussion of the ability of Americans to recognize, or think they recognize, African American and white speakers, even when they cannot see them. As I explore the construct of *ethnicity* as a tool for the investigation of the nexus of linguistic and social processes, I remain mindful of Rickford's observation that African American Vernacular English is "rooted in history, and . . . serve[s] to express the ideas, social relationships, and ethnic identities of thousands of African Americans daily" p. 274.

The volume also treats the encroachment of English on indigenous languages. Ortiz offers a moving first person narrative: "English has pushed Indigenous languages out of the Indigenous family, culture, and community, and this has brought about inevitable conflicts that run the gamut of intra-family relationships, tribal governance, and education. This problem and conflict has resulted in damming the flow of cultural and community continuity" p. 165. Yet, even in his account, class issues emerge: "We, Indigenous peoples of Acoma and Laguna Pueblos and the Navajos of Prewitt and Bluewater, were low income or no income poor people, simply cheap labor, who didn't seem to have much choice. No longer self-sufficient subsistence farmers, numbers of us went to work in the uranium industry. We were laborers for the most part or lower echelon skilled workers, never anything in management" p. 164. Patrick, treating many of the same issues, finds that for the indigenous Nunavut communities in Canada, English plays a "paradoxical role," "having the status of a language of colonization and dominance, yet at the same time serving as a necessary tool for the assertion of Inuit land rights and autonomy and for the protection of Aboriginal languages, rights, and local institutional control within the Canadian state" p. 168.

In one respect, it might appear surprising that a symposium convened to consider the theme of *English and Ethnicity* should have so much to say about other categories of social experience, including gender, class, race, and sexual orientation. For with regard to language and ethnicity, the focus, at least recently, has been, as Patrick notes, on the subject of language endangerment. The recent upsurge of interest in the topic has had as its aim, in part, to call the attention of linguists, scholars, policy makers, and the public to the potential disappearance of a large proportion of the world's linguistic, and therefore cultural, heritage (Bradley and Bradley 2002; Crystal 2000; Dalby 2002; Muhlhausler 1996; Nettle and Romaine 2000). Conveyed in discourse often borrowed from the environmentalist movement, it seeks to draw linguists into the political arena to defend the objects of their study—languages (cf. Mufwene 2002, 2004). There is, of course, need for linguists, like all scholars, to play a role in public policy and discourse in addition to their traditional academic pursuits, as Mary Louise Pratt (2003) cogently argues. But there are potential pitfalls as well. In entering the realm of political discourse, in which finer points of theory may appear remote and obscure, the complexities of linguistic processes can easily be lost sight of in the attempt to fit the case of language endangerment into the established categories of environmentalist political discourse.

Within this literature, English, and implicitly its relation to ethnicity, has been the most politicized subject of all. Consider, for example, the claim put forward in the introduction to a recent volume entitled *Language Endangerment and Language Maintenance*: "The globalisation of English and the spread of other national languages are not so different from the spread of new genetically modified plant varieties controlled by multinational companies" (Bradley and Bradley 2002, xii). Phillipson (2003) prefers the metaphor of the narcotics trade: "English has acquired a narcotic power in many parts of the world, an addiction that has long-term consequences that are far from clear. As with the drugs trade, in its legal and illegal branches, there are major commercial interests involved in the global English language industry" (16). After all, the rise of World English (Brutt-Griffler 2002) and the decline of a large proportion of the world's languages have largely coincided. Why not make the link direct? Nettle and Romaine (2000) do just that in *Vanishing Voices*: "Some have used the terms 'language murder' and 'language suicide,' suggesting that languages do not die natural deaths. They are instead murdered. English, as Glanville Price puts it, is a 'killer language' " (5). To bolster this point, they cite the case of Africa, in which English spread, they assert, "is leading to the top–down displacement of numerous other tongues" (144).

It is not a great step to go from English as the "killer language," the embodiment par excellence of "linguistic imperialism" (Phillipson 1992), to English as destructive of ethnic diversity. Phillipson helped popularize the prevalent notion in Applied Linguistics that learning a language means necessarily adopting an "alien" culture, becoming a victim of cultural as well as linguistic imperialism. Dorian (1999) reflects a common assessment of the inextricable "links between an ethnic group and its language" (31). More than simply "an identity marker," its "deeper connection" lies in the "extensive cultural content" it "carries" (31). The language encodes the group's history and the "people's sense of themselves" (32). "Core spiritual concepts framed in the heritage language of the group can be difficult or impossible to express with equal clarity or depth of meaning in another tongue. Much of this clarity or depth is inescapably diminished or lost when a people replaces its ancestral language with another" (32). According to Phillipson (2003), "the advance of English, while serving the cause of international communication relatively well, and often bringing success to its users, can represent a threat to other languages and cultures" (6), even portending, perhaps, language attrition and "a loss of cultural vitality" (176). For, again as represented by Phillipson (1999), it is taken for granted that only certain ethnic identities are authentically conveyed in English: "A speaker of English as a mother tongue may have one of several possible ethnic identities, Australian, British, Canadian, and so forth" (102). (The essays in this volume, particularly Rickford's, remind us that the picture is considerably more complex than that.)

Linguistics has long distinguished itself as a branch of science from the older grammarian tradition by characterizing the latter as *prescriptive* and itself as *descriptive*. And yet when we examine the literature pertaining to ethnicity we find, as the above examples illustrate, a pronounced tendency toward *prescriptivism* in the form of political advocacy—in notable contrast to the body of work in the fields of anthropology and sociology (Banks 1996). Perhaps those scholars at the forefront of this trend conceive that the term *applied* that they attach to the front of their discipline entails, despite Widdowson's (2003) injunction, the duty—or at least the space—to *prescribe*. Whatever the explanation, the complexities brought out by the essays in this volume demonstrate that linguists must be careful about embracing a disciplinary political orthodoxy that interprets *de-ethnicization* as the effect of cultural imperialism and ignores both the agencies and the instrumentalities involved. In the interests of enlightening what has been a rather thinly analyzed political discourse, there is the need for thorough investigation of the relevant linguistic processes at work that have shaped and are shaping language use in the world today. The essays that comprise this volume provide a window into those processes and therefore a useful entry point for considering them.

Construction of Ethnicity: Transformations of Race

In the "Introduction," Davies writes of her motivation to focus this volume on the topic of English and *ethnicity* in the interests of "resist[ing]" the use of *race* as an organizing analytical tool: "While linguists recognize that it is important to interrogate the naturalized notion of 'race' and to deconstruct it, . . . they are highly sensitive to the power of language to reify concepts" p. 6. And that provided my interest in exploring the direction the volume takes: it offers the opportunity, to use Davies's apt phrase, "to interrogate the naturalized notion" of *ethnicity* "and to deconstruct" its meaning and use in the field of linguistics (2).

The term *ethnicity* is of surprisingly recent vintage, dating back only to the 1960s (Banks 1996). Lacking equivalents in most of the world's languages, its use is largely confined to English-language contexts (Fishman 1999). And even within these, its meaning has proven remarkably elusive. Only one author in the present volume, Bernstein, explicitly sets out to define *ethnic group*, citing the National Council for the Social Studies' definition, one symptomatic of the lack of theoretical clarity that attaches to the notion. If *ethnic group* is, as that organization defines it, simply a cover term for groups demarcated by race, national origin, *or* culture, it would seem that in aspiring to include so much, the term might fail to signify much of anything at all. We might then be prompted to concur with Banks, an anthropologist whose survey of its use in the different disciplines has led him to conclude, "I do not think that ethnicity is simply a quality of groups, and for the most part I tend to treat it as an analytical tool, devised and used by academics" (4).

If so, it is one used by them for a multitude of purposes. For some, it provides a more attractive substitute for the discourse of race, particularly in the United States, where ethnicity is constructed largely on that basis. The call for the symposium that this volume grew out of repeats a formula that has gained some popular, though little scholarly, credence, in declaring, "the notion of ethnicity must be conceptualized as both subsuming and transcending earlier notions of 'race.' " Yet, in the determination of ethnicity in the United States, the racial other—that other of others in Anglo-American society—trumps all else. Mazrui reminds us that the American way of viewing ethnicity differs significantly from that found in other parts of the world. He remarks, "Any person who speaks Arabic as a first language could, in principle, claim Arab ethnic affiliation" p. 51. In contrast, although (as Rickford shows) African American culture has strong roots in Africa and the Caribbean, it remains true that "African Americans could not associate themselves with the dominant Anglo-American identity simply by virtue of being 'native' speakers of English" p. 51. On the contrary, as already noted, American Africans become ethnically African American (whatever their specific national origin or culture) "at precisely the point when they lose their ancestral languages and acquire the English language instead" p. 69.

It might seem reasonable, even enlightened, to claim that ethnicity supersedes "older notions of race." But racialized societies like the United States have manifested the tendency for ethnicity to devolve into race. Consider the term "ethnic American." Logically, it should, following the formula used in other cases (e.g., "ethnic Serb"), denote a person of American ethnicity. Yet, the term is more commonly used to indicate a nonwhite person[2]—with the most popular designations being—African American, Asian American, Hispanic or Latino American, and Native American. Farr explicitly calls attention to the degree to which Hispanic has become a racial term. Toribio neatly illustrates the point as well, since for Dominicans becoming a Standard English monolingual removes the one exit from an African

American identity—the maintenance of a Hispanic one. Asian American groups together East and South Asians, despite the absence of any meaningful common heritage. Moreover, if we substitute persons of European origin for those of African, Mazrui's statement quoted above would read: "European Americans can avoid an ethnic identity simply by virtue of being 'native' speakers of Standard English." If Jewish American (or Polish or Italian American) is an ethnic identity, then white American would seem not to be. White Americans' ethnic identity, like that of "ethnic Americans," is replaced by a seemingly racial one. It has been argued that they do not have any socially constructed ethnic identity at all. Sociologists have adopted the term *optional* or *symbolic* ethnicity to describe the situation of Americans of European descent—who can choose to either claim a particular European ethnicity or remain outside ethnic classification (Waters 2001). Waters notes, "For all of the ways in which ethnicity does not matter for White Americans, it does matter for non-Whites . . . whose lives are strongly influenced by their race or national origin regardless of how much they may choose not to identify themselves in terms of their ancestors" (432).

Waters exposes the weakness of the notion that, as Banks (1996) describes it without subscribing to it, "properly understood, ethnicity subsumes race" (51). On the contrary, for race to be an aspect of ethnic identity, ethnicity must be racially bounded, or at least potentially so. Claiming that ethnicity "subsumes older notions of race" means taking race as a building block of ethnic identity. Such a conception thereby *constructs ethnicity on the basis of race.* In contrast, the converse is not true, at least in the United States, where ethnicity is largely ignored in the construction of race. For example, Africans are classified as black, and Europeans as white when they assimilate into North American culture regardless of their actual ancestry. As Mazrui points out, there is no recognition of Nigerian American or Angolan American even for the children of immigrants from those nations; and Farr notes that Mexicans who come to the United States tend to be subsumed under the category of Latino or Hispanic, designations "contested by these Mexicans both implicitly and explicitly" p. 244. That is just as evident in the case of European American identities insofar as they are optional— for the group into which persons of European origin disappear is a *racial* category. Insofar as African American, Hispanic American, Asian American, Native American, and European American (white) are ethnic categories, ethnicity is *transformed into race.* Ethnicity, then, only "subsumes" race by replicating it, by dissolving into it.

In fact, the same case that Davies makes in the introduction to this volume for avoidance of the discourse of race can be made for dropping the discourse of ethnicity. Waters writes,

> There is a tendency to view valuing diversity in a pluralist environment as equating all groups. The symbolic ethnic tends to think that all groups are equal; everyone has a background that is their right to celebrate and pass on to their children. This leads to the conclusion that all identities are equal and all identities are in some sense interchangeable— "I'm Italian American, you're Polish American. I'm Irish American, you're African American." The important thing is to treat people as individuals and all equally. However, this assumption ignores the very big difference between an individualistic ethnic identity and a socially enforced and imposed racial identity. (432–433)

The point that some scholars seem to overlook in their narrow focus on identity is that if ethnicity is socially rather than individually constructed, then it must also be socially rather than individually recognized. That is, not all ethnic identities are optional or elective, nor are they rooted in identities that serve the individual to which they are socially ascribed. The transitions from, say, Chinese to Asian American, from Mexican to Hispanic American, from Nigerian to American African (or African American) constitute not mere changes of ethnic identity, but rather the transformation of an ethnic identity into one based on socially constructed notions of race. And the transition from, say, German to white American is not so much a change in ethnic identity as the discarding of one altogether, perhaps supplemented by the maintenance of the optional, or symbolic, identity of German American. Ethnicity no longer appears as a neutral social identity common to all.

The notion that unlike the discourse of race that of ethnicity is somehow benign ignores the instrumentalities of ethnicity—and so it is no surprise that race constitutes perhaps the dominant theme throughout this volume. To argue, therefore, as Davies does in the introduction to this volume, that to substitute the discourse of ethnicity for that of race is to be "highly sensitive . . . to the power of language to reify concepts" (2) ignores the specific content of ethnicity as racial discourse under another name, at least in the United States, where the tendency is that *ethnicity signifies race.*

Critically analyzed, there is little to prefer in the discourse of *ethnicity* as opposed to that of *race.* For the former entails (like the notion of race) the distinguishing of group members from nonmembers in the larger society via *construction* and *exclusion* of a social *other,* whether on racial, national, cultural, religious, or linguistic grounds. In the abstract, ethnic groups, since they can be socially identified, be it through what Coupland terms *acts of identity* or by other means, might appear to represent coherent and cohesive social groupings (as for that matter, might racial groups). There are, however, as Johnstone (2000) argues, significant conceptual problems with such an approach:

> We should also be aware, as Cameron et al (1992) point out, that research that *studies* groups of people also has the effect of *creating* groups of people. No matter how many disclaimers are added, labeling a linguistic variety "African-American Vernacular English" or "Southern speech" creates groups of speakers, African-Americans or Southerners, potentially obscuring the fact that African-Americans and Southerners have many identities and many ways of talking. People do not like to be told that they act the way they do because of social facts about them, and we should take this seriously. To suggest that people's behavior is determined by their group memberships, as sociolinguists often do, is to suggest that people do not have individual voices and do not make creative, responsible choices, and thus to deny people an aspect of their humanity. (54–55)

This is all the more the case since, in practice, scholarly constructions of ethnicity as, in Banks's terms, "an analytical tool, devised and used by academics," *re*construct existent communities on an exclusionary basis. As Chapman et al. (1989, quoted in Banks) comment: "[Ethnicity] is a term that half-heartedly aspires to describe phenomena that involve everybody, and that nevertheless has settled in the vocabulary as a marker of strangeness and unfamiliarity" (4). Coupland notes that ethnic identity is centrally concerned with "the sorts of boundary work that people do." Such boundaries, however, take root in *exclusion of the other*. For instance, Farr writes, "*Rancheros* maintain racial boundaries between themselves and the indigenous in both linguistic and non-linguistic ways. Language is used to distinguish the indigenous either through their use of Purhepecha or through the way they speak (their dialect) of Spanish" p. 238. The same underlying racialization of language use appears in Torribio's study.

NARRATIVES OF ETHNICITY: THE ANCESTRY MYTH

The *uncritical* adoption for scholarly purposes of constructs such as *race* and *ethnicity* runs into, in addition to their endemic contradictions, the myths with which they are inextricably bound. In his essay, Huang notes that Chinglish as a "translocal dialect" p. 76 "not only transcends geographical boundaries, but also unsettles the putative connection between a dialect, and a localized, romanticized origin" p. 96. Here Huang gets to the heart of the uses of the past that permeate questions of language and culture. Indeed, when linguists lament that the loss of language represents a loss of culture, they unconsciously shift from a view of culture as process to one of culture as artifact (cf. Mufwene 2004). It is forgotten that, for instance, in declaring, as Dorian (1999) does, "Although many behaviors can mark identity, language is the only one that actually carries extensive cultural content" (31), that such a statement holds true for *any* language. Not only, then, can "a culture" express itself in a "new" language—it *must* do so. If language "encodes human experience" (32), then any language we speak encodes our experience—as Rickford's essay illustrates, as does the existence of the New Englishes (Kachru 1992). To give precedence to certain expressions of our experience over others is a normative approach, a value judgment—a privileging of past over present experience. It is as though as soon as the topic shifts to disappearing languages, some linguists suddenly forget all their own views on socially constructed meaning, and revert to an essentialist, static view of language and culture. Mufwene (2002) has, therefore, rightly noted, "It is perhaps important . . . that we in linguistics learn the distinction between preserving a language (like a museum piece), maintaining it in usage, and revitalizing it (by restoring vitality to it). Realistically, we have more control over preservation than over maintenance and revitalization" p. 29.

There is an inherent and as yet largely unarticulated conceptual flaw that attaches to ethnicity as an analytical category. For it implicitly attempts to draw an intrinsic connection between two incommensurate qualities—ancestry and culture. The transmission of the languages and cultures with which linguists such as Dorian concern themselves is implicitly or explicitly held to be the task of ethnic groups. To construct ethnicity on the basis of alleged ancestry— "*belief* in a common ancestry" (Waters 2001, 430–431) is the most typical formula—is to do so on the basis of very subjective uses of the past. Farr's field research clearly shows such a process: "these *rancheros*, like others in pockets all over western Mexico, construct themselves as non-indigenous, even while acknowledging their *mestizaje*" p. 232. Such a construction of their ancestry represents a rejection of part of their heritage in favor of another part. After all, no one actually knows his or her own ancestry with any degree of certainty, all the less anyone else's. To attempt to instantiate ancestry in a world whose history has consisted of constant migrations, dislocations, and geographical fluidity manifests a tendency to shift the focus from history as individual and social *lived experience* to *history as myth*.

And yet, the literature on ethnicity is replete with assertions like the following: "a person can be institutionally naturalized as a national, whereas one still has to be born into ethnicity" (James 1997, quoted in Fishman 1999, 447) and, "You are born into a specific ethnic group" (Skutnabb-Kangas 1999, 55). In the first place, of course, such notions constitute vast oversimplifications of the processes that, for example, render ethnicity in the United States. Clearly, for instance, a person born in Mexico is not a Hispanic American, nor a person born in China an Asian American, as they become on settling in the United States. And what about the African American children of American Africans? Their ethnic identity does not at all follow that of their parents. Nor are such ethnicities universally recognized. Padilla (1999), for example, notes that, though he is identified as "Hispanic" in America, "In Latin America, I have been identified as an American who speaks fairly good Spanish" (118–119). In the same way, neither Asian nor African represent ethnic categories on those continents. Nor does it particularly enlighten the discussion to simply declare, as Fishman does, "our ethnic identity changes from one occasion to another" (153). Such observations do little to uncover the nature of ethnicity, its origin, or the social purposes it serves. As much as some researchers have tried to decouple the notion of ethnicity from that of ancestry or birth, the socially constructed notion from which it is derived stubbornly resists such attempts, as demonstrated by the racial substrate to ethnicity in the United States.

The problem is in some ways even more fundamental. Analysis rooted in notions of ethnicity consists in the attempt to impose uniformity on a world of hybridity—to reify at least partially imaginary notions. For instance, one of the most prolific proponents of viewing the world through ethnically essentialist terms, Skutnabb-Kangas, writes, "Both ethnicity and an attachment to one's language or mother tongue(s) as a central cultural core value seem to draw on primordial, ascribed sources: You are born into a specific ethnic group, and this circumstance decides what your mother tongue (or tongues, if your parents speak different languages) will initially be" (55). And what happens when a person with two mother tongues marries someone with two others? Will their children have four? Despite Skutnabb-Kangas's hedging, ethnic analysis relies on the notion that the boundaries between ethnic groups involved in the "boundary work" to which Coupland refers remain viable and *real*. In multi-ethnic settings, however people would have to marry within their ethnicity for such boundaries to be maintained, and yet this fails to occur in enough cases that ethnic identities begin to break down. Mufwene (2004) notes of the African case, "The gradual obliteration of ethnic boundaries, caused in part by interethnic marriages, has been an important factor in the loss of ethnic languages" p. 212.

The essays in this volume serve to illustrate that throughout the world alongside the forces that work to maintain ethnic consciousness there are equally significant processes of *de-ethnicization*, the exit from ethnicity discussed in Mazrui. Yet where they do occur, and involve groups of non-European origin, there is a tendency to ascribe them to Western agency, via the ubiquitous explanation of cultural imperialism, as in Africa. Phillipson (1999), for example, in explaining the disfavor for notions of ethnicity among black South Africans, attributes it to the apartheid past. He also suggests that it would be wrong to conclude that the "wellsprings of ethnic identification do not exist among South Africans" since "the protracted dormancy of ethnic identifications in many parts of the world, Communist and capitalist, was misleading" (104). Phillipson implicates the learning of English in "postethnicity" (104) and contrasts it with the use of African languages, which, he implies, carry the "wellsprings of ethnic identification." In common with the mainstream of the language rights and language endangerment literature, he wrongly assumes that English currently represents the greatest threat to "indigenous" languages in Africa. Such a view ignores the emergence of "indigenous" African lingua francas—across the continent, particularly in urban areas (Makoni and Brutt-Griffler 2007; Mazrui 2004; Mufwene 2004; Winford 2003). These mixed languages, like South Africa's Isicamtho, not English, are replacing ethnic languages as the mother tongues, or languages of primary socialization, of increasing numbers of speakers. South African poet Ike Mboneni Muila (2004) describes Isicamtho as "a language which draws from and brings together all South African languages that kept people apart" and "a new profound language . . . a language of identity . . . unity in diversity."

And it points out another social process lost in the abstract terms in which scholars such as Phillipson and Skutnabb-Kangas frame the question: over the long term, children tend to acquire their mother tongues *not from their parents*, but *from the society* in which they are socialized. This condition is demonstrated in South Africa by the switch to language socialization in urban vernaculars, rather than the institutionally recognized mother tongues (Childs 1997; Makoni and Brutt-Griffler, 2007). But it is also suggested by Canagarajah's study—otherwise the children of Tamil-speaking parents could not become English monolinguals. The ethnic group is by its nature an unstable category of social existence, subject to constant disruption by powerful social processes, including migration. At the core of the belief in the meaningfulness of ethnicity as a category is the notion that it *travels*: that the German who moves to Poland remains German, or a Tamil who relocates to Canada remains Tamil, ostensibly for all time. Once again it attempts, through ancestry, to tie individuals to points in space and time (or myths of places and histories)—and via these to cultures—that might lie outside the lived experience of those who are held to be its carriers. At best, it presents individuals with a dilemma—the type that Canagarajah's informants convey in the ambiguity they express toward their "social identity." At the extreme, it attempts to tie persons narratively to what may already represent a completely unfamiliar, even alien, culture. And yet, conceived, generally for racial reasons, as the ethnic other, they may be bounded off from claiming membership in the only "ethnic culture" they know.

This type of thinking is not without practical implications. It translates into the architecture of the primary area of social policy-making designed to uphold ethnicity—language policy. The flaws of the abstract terms in which Skutnabb-Kangas and Phillipson frame the question come through in their prescriptive statements on the optimal construction of language policy. Phillipson (1999) writes (adopting Skutnabb-Kangas's formula): "The challenge of reducing English [in South Africa] to equality involves ensuring that English is learned as an additional language" (105). Although it is formulated as a question of the rights of a language (as though rights were not a political doctrine pertaining to persons and not social phenomena like languages), its consequences all fall on persons. How is the goal in question to be accomplished except by insisting on the maintenance of mother tongues, and ignoring the individual rights of persons to learn whichever language they might chose? The problem is with the intention Phillipson expresses of "ensuring" such a result. The implication is that there is no, or should be no, exit from ethnicity as a matter of state policy. Phillipson's collaborator, Skutnabb-Kangas (1999), argues for something approaching such a notion even more directly: "I do not agree with those researchers who see both ethnicity and a mother tongue in an instrumental way, as something you can choose to have or to not have, to use or not use, according to your own whims and wishes. Because of the primordial sources that reach back into infancy and personal history, neither ethnicity nor mother tongue nor even identities can be treated as things, commodities, that one can choose and discard like an old coat at will" (55). According to Skutnabb-Kangas, a given individual's "whims and wishes"—their own free will—should not enter into the question. The best means of preventing such choice is to *enforce* via policies of the type they advocate that only certain mother tongues, and with them certain ethnic identities, are made available by the educational system of a nation like South Africa. "You are born into a specific ethnic group, and this circumstance decides what your mother tongue (or tongues, if your parents speak different languages) will initially be" (55). Of course, if you cannot "discard [it] like an old coat," the adverb *initially* is misleading.

Such prohibitions against those of certain ethnicities learning English as a mother tongue—"natively"—ensures that, as second language users, they will be set off very effectively by one of the crucial linguistic signs of exclusion—nonnative proficiency in the language. As Phillipson (1999) notes, as a mother tongue speaker, an individual can qualify for certain ethnic affiliations. Without it, they cannot. Is the state really to be allowed into the home to dictate which language is used there in the name of "ensuring" that English is "reduced to equality"? Such abstract formulations, rather, simply result in cutting off choices via language policies specifying that only certain students should have access to English-medium schools. Such policies simultaneously put the state in the business of legislating ethnic identities in shutting individuals off from the "native" proficiency in a language like English that enables *exit from ethnicity*. The problem is, once we begin to safeguard ethnicities, it is very easy to turn the whole process on its head. Protecting Black South Africans from the alleged "imposition" of a mother tongue English identity involves precisely the same measures as protecting that identity from black persons. As long as the English spoken is accented, nonnative, non–mother tongue, then language serves as a *sign of exclusion*.

The point that Skutnabb-Kangas and Phillipson miss is that the shifts in ethnic identity that take place represent complex social processes, not simply individual choices (although these are of course also involved). The African urban experience demonstrates the danger inherent in linguists adopting a political orthodoxy that sees the world in overly simplistic, binary terms. That the *ancestor myth* is seldom interrogated to reveal the sorts of hegemonic or ideological forces that come through in, for example, Farr's account says much about the purposes that the discourse of ethnicity is used to serve.

THE INSTRUMENTALITY OF ETHNICITY: THE POLITICS OF EXCLUSION

The uncritical adoption of ethnicity as a category drawn from experience leaves another question unanswered, that of *why* people in society should identify with certain people to the exclusion of others. To simply declare that it is enough to observe that they do (Phillipson 1999; Skutnabb-Kangas 1999) begs the question. The answer lies not only in the use of history as a subjective, ultimately emotional, force, but also involves its usage as such for certain instrumental reasons. The social construction of ethnicity can be made to serve purposes of what Phillipson calls "hegemonic ordering" (if we accept such a problematic term) just as easily as the "cultural imperialism" that is alleged to threaten cultural diversity. To return to Farr's ethnography of the *rancheros* who "construct themselves as non-indigenous," p. 232 it cannot be lost sight of that such a self-conception serves the interests of upholding a "racial ideology that values light skin" p. 233. These cases of *acts of identity* serve explicitly *instrumental* purposes, a point that sociologists and political scientists have drawn attention to (Banks 1996; Phillipson 1999), and that linguists have largely overlooked. We hear much these days in Applied Linguistics about the threat of cultural and linguistic imperialism, little about the force of racism and national and ethnic chauvinism, despite their presence in the very discourse that makes up the foundation of the field (Brutt-Griffler and Samimy 2001). It is as though the forces of *de-ethnicization* are purely ideological, whereas those upholding *ethnicity* are removed from the world of ideology, hegemony, and other means of social control.

If the call of "ancestry" is to overcome the pull of the immediate circumstances in which one might live, there is the need to overcome the pragmatic necessities of immediate circumstances. As one of Canagarajah's informants expresses it, "This is London, not Jaffna. What's the point of speaking in Tamil?" p. 195 The circumstances of language maintenance are just as politicized as those of language spread, and just as likely to serve nefarious political and economic interests—and it is just as incumbent on linguists to study and call attention to them. Mazrui's work provides an excellent case in point. He shows that the call to ancestry over pragmatic necessities may exploit forms of social oppression involving not only race but gender and class as well. He notes, in "the more 'traditional' rural setting . . . women are regarded as the custodians of ethnic culture," a role that entails "their subordinate status within the ethnic community" p. 66.

The notion of ethnicity, then, may involve the exertion of social control through ideological means—a "hegemonic" project. Moreover, it is one easily bent to overtly political purposes. As Wilmsen and McAllister (1996) write, "individuals are persuaded of the need to confirm a sense of identity in the face of threatening economic, political, or other social forces" (ix). Ethnic division differentiates, by a process of exclusion, the other, recreating the "us" versus "them" consciousness that lies at the heart of social conflict. And, of course, ethnicity has been heavily implicated in recent wars, and even "ethnic cleansing" (Banks 1996; Hughey 1998), something Fishman (1999) notes but claims to be as true of "the other manifold memberships that may be affirmed in times of need, be these political party, gender, age, religious, class or any other memberships" (447).

It is, then, all the more significant that those experiencing nonethnic forms of oppression within an ethnicized society may be the first to seek exit from ethnicity—as suggested in both Mazrui and Canagarajah. Mazrui notes that though English is withheld from Zulu women in rural South Africa, in urban areas they are "more highly motivated to learn the language because it accord[s] them new opportunities to escape from their ethnically ascribed status on the grounds of their gender" p. 66. My field research in Cape Town, South Africa, shows that the motivation of young females to learn English also involves issues of class. In the words of one informant, "You go to school [in the black township], you don't finish school, you get pregnant sometimes, you have to stay and help your mother, or you work as a domestic worker— your parents will ask you to go find a job, if you don't know English, so you work as a domestic worker" (Brutt-Griffler 2005). In contrast to the attention given over to the analysis of the economic interests involved in globalization, there is almost none given to how the politics of ethnic identity may serve the interests of the local elite to maintain class privilege and socioeconomic power (Brutt-Griffler 2005).

Given the close connection of notions of ethnicity to such hegemonic ideological forces as racism and ethnic chauvinism, we might well be led to wonder why processes of *de-ethnicization*, of the kind evidenced in Mazrui and Canagarajah, do not receive equal attention as those representing ethnic identities upheld, or why when they are referenced they are so often held to be threatening to the world in which we live (or want to live). Toribio goes a significant way toward identifying such processes:

> The interviews with the youths further exposed, rather suggestively, that just as their parents' language practices appear to be fading, so too

do their parents' racial attitudes. For unlike first-generation immigrants who formulated their impressions of African Americans by observing them at a distance, the second generation takes in additional data with which to construct a more informed view.

In summary, rather than seeking expressions of status and prestige, as defined in the U.S. setting, the majority of the adolescents interviewed appear to manifest a solidarity with their black and white peers and a new discourse of intimacy with their compatriots. For these youths, "being" Dominican in the diaspora extends beyond the application of self-label for self-categorization to the communication of a new, more inclusive Dominican narrative. In doing so, they advance towards dismantling essentialist concepts of Dominican identity (as nonblack, Spanish-speaking, etc.).

Toribio's pioneering work brings out nuances of this process that have been ignored by the static terms in which sociolinguists have often viewed ethnic identity.

FUTURE DIRECTIONS

As the analysis above reveals, just as racism serves as the motivation—at least historically speaking—for identifying others by race (cf. Taylor 2005), so too are racism, nationalism, and xenophobia intimately connected with the socially constructed notion of ethnicity. My intention in this conclusion is not to suggest that social scientists can dispense with *ethnicity* as an analytical tool, however imprecise, any more than we can afford to cease to investigate social phenomena related to race. On the contrary, we cannot change the world by *failing* to analyze it. Like the discourse of race, that of ethnicity cannot be taken as a substitute that places all social groups on an equal footing. Scholars, including linguists, must investigate all of its social manifestations and uses, and not content themselves with mapping its social geography and charting its distinguishing features.

As the essays in this volume illustrate, race is too intertwined with the notion of ethnicity to allow the study of the latter in the absence of the former, providing elegant justification for the inclusion of the volume *English and Ethnicity* in the series "Signs of Race." Hence, the formula expressed in the symposium call—"Our theoretical position is that ethnicity is *potentially* an aspect of the identity of every person, and that English can be used to signal a wide range of ethnicities in a wide range of social contexts" p. 2—requires modification. As follows from Mazrui's discussion, in racialized contexts like the United States, English can be used to signal ethnic identity only *in combination with race*. It confers, in this case, an American identity, but only as a hyphenated one (African-American, Asian-American, etc.), with the hyphen representing an ethnic *sign of race*.

The body of work from the volume highlights the importance of multiple approaches to issues of language and ethnicity, as shown by the many useful insights that emerge from outside the traditional socio-linguistic focus on language variation. Methods rooted in a rapidly receding past provide only one window into a fundamentally altered English-speaking world. The symposium, for instance, turned up little evidence of a link between English variation and *optional ethnicity* among white Americans, Bernstein's essay being the exception. And, of course, since race underlies nonoptional ethnicity, the importance of language use in signaling nonwhite ethnic identities is necessarily limited. On the contrary, the volume demonstrates that ethnicity in many cases, perhaps the majority of instances involving the use of English as a mother tongue in the United States at least, is not sig-naled linguistically at all, but derives from quite different sources. To that end, sociolinguistics needs to chart the degree to which language and purported ethnic identities *diverge*—the degree to which varieties of English do *not* signal ethnic identities as socially constructed—by paying greater attention to the ways in which language use does not uniformly index these identities. It also needs to broaden considerably beyond the assumptions that have tended to underlie variationist linguistics to find windows into social processes of de-ethnicization. We live in an increasingly transnational world, a reality that is not always sufficiently reflected within sociolinguistics, which has main-tained to too large an extent a focus on intranational variation, despite the increasing disappearance of traditional categories of analysis (regional and class variation).

Even in the United States, *bilingualism* is already challenging both regional and class variation within English usage in its significance as an identity marker. Studies like Toribio's and Canagarajah's take the field forward in new and interesting ways, showing that the relation between national origin and bilingualism may take surprising and hitherto unsuspected forms. (On new directions in bilingualism, see Brutt-Griffler and Varghese 2004.) From a global perspective, within a language with a majority of non–mother tongue users, the con-struction of ethnic identities will also tend to be signaled more by bilingualism, including but not limited to the use of English as a non-native language, than by readily socially identifiable (monolingual) ethnic language varieties. Hence, at the heart of the sociolinguistic investigation of English moving forward will increasingly be the inter-related questions of World English and bilingualism, as the demo-graphics of English speakers in the world shifts toward the further dominance of these categories of speakers.

NOTES

1. For a more detailed treatment of these issues, see Mazrui (2004).
2. Interestingly, this is the sense in which the term is applied by those authors who reference ethnicity to the American context in the volume *Handbook of Language and Ethnic Identity* (Fishman 1999). Padilla (1999) uses the term "ethnic people in America" as synonymous with "minorities" (119). Bourhis and Marshall (1999) use "ethnic populations" (245) as an equiv-alent for the same term.

REFERENCES

Banks, Marcus. 1996. *Ethnicity: Anthropological Constructions*. New York: Routledge.

Bourhis, Richard Y. and David F. Marshall. 1999. The United States and Canada. In *Handbook of Language and Ethnicity*, J. Fishman (ed.). New York: Oxford University Press. 244–267.

Bradley, David and Maya Bradley (eds.). 2002. *Language Endangerment and Language Maintenance*. London: Routledge Curzon.

Brutt-Griffler, Janina. 2002. *World English. A Study of Its Development*. Clevedon: Multilingual Matters Press.

Brutt-Griffler, Janina. 2005. "Who do you think you are, where do you think you are?": Language policy and the political economy of English in South Africa. In *The Globalisation of English and the English Language Classroom*, Claus Gnutzmann and Frauke Intemann (eds.). Tübingen: Narr. 25–37.

Brutt-Griffler, Janina and Keiko K. Samimy. 2001. Transcending the native-ness paradigm. *World Englishes* 20, 1: 99–106.

Brutt-Griffler, Janina and Manka Varghese (eds.). 2004. *Language Pedagogy and Bilingualism*. Clevedon: Multilingual Matters Press.

Chapman, Malcolm, Maryon McDonald, and Elizabeth Tonkin. 1989. Introduction. In *History and Ethnicity*, Elizabeth Tonkin, Maryon McDonald, and Malsolm Chapman (eds.). New York: Routledge.

Childs, G. Tucker. 1997. The status of Isicamatho, an Nguni-based urban variety of Soweto. In A.K. Spears and D. Winford (eds.). *The Structure and Status of Pidgins and Creoles*. Amsterdam and Philadelphia, PA: John-Benjamins. 341–370.

Crystal, David. 2000. *Language Death*. New York: Cambridge University Press.

Dalby, Andrew. 2002. *Language in Danger*. New York: The Penguin Press.

Dorian, Nancy C. 1999. Linguistic and ethnographic fieldwork. In *Handbook of Language and Ethnic Identity*, J. Fishman (ed.). New York: Oxford University Press. 25–42.

Fishman, Joshua (ed.). 1999. *Handbook of Language and Ethnic Identity*. New York: Oxford University Press.

Hughey, Michael (ed.). 1998. *New Tribalisms: The Resurgence of Race and Ethnicity*. New York: New York University Press.

James, Paul. 1997. *Nation Formation: Towards a Theory of Abstract Community*. London: Sage.

Johnstone, Barbara. 2000. *Qualitative Methods in Sociolinguistics*. New York: Oxford University Press.

Kachru, Braj. 1992. *The Other Tongue*. Chicago: The University of Chicago Press.

Makoni, Sinfree, Geneva Smitherman, Arnetha F. Ball, and Arthur K. Spears (eds.). 2003. *Black Linguistics: Language, Society, and Politics in Africa and the Americas*. New York: Routledge.

Makoni, Sinfree, Brutt-Griffler, Janina and Pedzisai, Mashiri. 2007. The use of "indigenous" and urban vernaculars in Zimbabwe. *Language in Society* 36(1).

Mazrui, Alamin M. 2004. *English in Africa: After the Cold War*. Clevedon, UK: Multilingual Matters Press.

Mufwene, Salikoko. 2002. *Colonisation, Globalisation, and the Future of Languages in the Twenty-First Century*. Available from World Wide Web: http://www.unesco.org/most/vl4n2mufwene.pdf.

Mufwene, Salikoko. 2004. Language birth and death. *The Annual Review of Anthropology* 33: 201–222.

Muhlhausler, Peter. 1996. *Linguistic Ecology: Language Change and Linguistic Imperialism in the Pacific Region*. New York: Routledge.

Muila, Ike Mboneni. 2004. http://www.donga.co.za/interviews/muilainterview.html.

Nettle, Daniel, and Suzanne Romaine. 2000. *Vanishing Voices: The Extinction of the World's Languages*. New York: Oxford University Press.

Padilla, Amado. 1999. Psychology. In *Handbook of Language and Ethnic Identity*. New York: Oxford University Press. 109–122.

Phillipson, Robert. 1992. *Linguistic Imperialism*. New York: Oxford University Press.

Phillipson, Robert. 1999. Political Science. In *Handbook of Language and Ethnic Identity*, J. Fishman (ed.). New York: Oxford University Press. 94–109.

Phillipson, Robert. 2003. *English-Only Europe? Challenging Language Policy*. New York: Routledge.

Pratt, Mary Luise. 2003. Building a New Public Idea about Language. *Profession 2003*: 110–119.

Skutnabb-Kangas, Tove. 1999. Education of Minorities. In *Handbook of Language and Ethnic Identity*, J. Fishman (ed.). New York: Oxford University Press. 42–60.

Taylor, Gary. 2005. *Buying Whiteness: Race, Culture and Identity from Columbus to Hiphop*. New York: Palgrave.

Trudgill, Peter. 1995. *Sociolinguistics: An Introduction to Language and Society*. New York: Penguin Books.

Waters, Mary. 2001. Optional ethnicities: For Whites only? In *Race, Class, and Gender: An Anthology*, 4th edition, M.L. Andersen, and P. Hill Collins (eds.). Belmont, CA: Wadsworth/Thomason Learning. 430–439.

Widdowson, Henry. 1994. The ownership of English. *TESOL Quarterly* 28, 2: 377–381.

Widdowson, Henry. 2003. *Defining Issues in English Language Teaching*. Oxford: Oxford University Press.

Wilmsen, Edwin N. and Patrick McAllister (eds.). 1996. *The Politics of Difference; Ethnic Premises in a World of Power*. Chicago: University of Chicago Press.

Winford, Donald. 2003. *An Introduction to Contact Linguistics*. Oxford: Blackwell.

INDEX

(Please note that page numbers in *italics* indicate an endnote.)